MULTIHAZARD RISK ATLAS OF MALDIVES

Geography—Volume I

MARCH 2020

Notes:
In this publication, "$" refers to United States dollars.
The maps presented in this atlas reflect airports based on 2017 data from the Civil Aviation Authority of Maldives.

On the cover: An aerial view shows 1 of 26 natural atolls that make up Maldives, which also includes nearly 1,200 small
coral islands and some of the world's most beautiful beaches. Recognized as the seventh-largest in the world, the coral
reefs and associated ecosystems of Maldives are key foundations for food security and means of livelihood. Yet, they are
considered as among the most vulnerable to climate change (photo by Roberta Gerpacio).

Contents

Tables and Maps

Tables

Maps

Foreword

Maldives is among the countries most vulnerable to the impacts of climate change as it is a small island nation with extremely low elevations. Maldives is also very vulnerable to impacts of rising air and sea surface temperatures and changes in rainfall patterns. Climate change impacts will therefore impose significant negative consequences on the Maldivian economy and society. Some of the priority vulnerabilities to climate change are land loss and beach erosion, infrastructure damage, degradation of coral reefs, and adverse impacts on water resources, food security, human health, and the overall economy.

Sustainable coastal resources management is of particular importance to Maldives, such that all regulations involving various development activities have coastal components. Despite the government's continued efforts in improving and sustaining coastal resources management, critical issues remain, such as the need for systematized coastal monitoring, clear definition of coastal boundaries and coastal development, enhanced regulatory and monitoring capacities for coastal resources protection, and sustainable long-term strategies on land reclamation and marine area protection. At a time when climate is rapidly changing and extreme weather events are frequently occurring, the critical roles that marine and coastal environments play in mitigating and adapting to climate change need to be sufficiently documented and properly recognized. It is therefore essential for Maldives to develop and establish a comprehensive digital database of marine and coastal ecosystem features and services that can be regularly monitored.

The *Multihazard Risk Atlas of Maldives* was developed through the project "Establishing a National Geospatial Database for Mainstreaming Climate Change Adaptation into Development Activities and Policies in Maldives" under the Asian Development Bank's regional knowledge and support (capacity development) technical assistance Action on Climate Change in South Asia (2013–2018). This five-volume atlas aims to promote the sustainable development of coastal and marine ecosystems and their various components, by enhancing the awareness of stakeholders on and enjoining them to address climate and disaster risks (including hazards, exposures, and vulnerabilities) to which ecosystems are exposed. The atlas presents spatial information and maps necessary for assessing future development investments in terms of their risks to climate and geophysical hazards.

The target audience of the *Multihazard Risk Atlas of Maldives* are the concerned stakeholders with current or planned development activities in the country, including public and private sectors, nongovernment organizations, research and academic community, development partner agencies, other financial institutions, and the general public. The atlas will also be a useful reference for other developing countries with similar geographical and environmental conditions, particularly small island developing states. It is envisioned that the atlas will significantly contribute to rendering important sector development investments more resilient to hazard-specific risk scenarios in the short, medium, and long terms.

H.E. Dr. Hussain Rasheed Hassan
Minister
Ministry of Environment, Malé

Shixin Chen
Vice-President for Operations 1
Asian Development Bank, Manila

Acknowledgments

Government Ministries, Departments, and Agencies in Maldives
- Civil Aviation Authority
- Land and Survey Authority
- Marine Research Institute
- Meteorological Service
- Ministry of Economic Development
- Ministry of Education
- Ministry of Environment
- Ministry of Fisheries, Marine Resources and Agriculture
- Ministry of Health
- Ministry of National Planning and Infrastructure
- Ministry of Tourism
- National Bureau of Statistics
- National Disaster Management Center

International Institutions
- Manila Observatory
- Marine Spatial Ecology Lab, University of Queensland, Australia
- SANDER + PARTNER
- United Nations Development Programme

International Institutions in Maldives
- International Union for Conservation of Nature, Maldives
- United Nations Development Programme, Maldives

National Consultant Team
- Ahmed Jameel, Integrated Coastal Zone Management Specialist
- Faruhath Jameel, Geographic Information Systems Specialist and Team Leader
- Hussain Naeem, Coastal Ecosystems and Biodiversity Specialist
- Mahmood Riyaz, Climate Change Risk Assessment Specialist

Abbreviations

BODC	–	British Oceanographic Data Centre
CAA	–	Maldives Civil Aviation Authority
GEBCO	–	General Bathymetric Chart of the Oceans
IHO	–	International Hydrographic Organization
IOC	–	Intergovernmental Oceanographic Commission
km^2	–	square kilometer
ME	–	Ministry of Environment
MED	–	Ministry of Economic Development
MLSA	–	Maldives Land and Survey Authority
MNPI	–	Ministry of National Planning and Infrastructure
UTM	–	Universal Transverse Mercator
WGS	–	World Geodetic System

Risk Mapping: Making the Invisible Visible

Maps have changed the way we see the world. By symbolizing the features of the earth and drawing its visible and invisible boundaries, maps give humans a wider perspective, which allows us to understand the patterns, trends, and interconnected components of our planet and see beyond where we have traveled. What used to be invisible to some became visible for all through maps.

At the global scale, the size of Maldives fades in comparison with its neighboring countries such as India and Sri Lanka. Sitting in the middle of the Indian Ocean, Maldives is barely visible but greatly vulnerable to natural hazardous elements, and it strives to overcome these challenges. People have lived in this beautiful nation for thousands of years despite limited land space and multiple hazards. However, their lives, livelihoods, and properties are becoming increasingly exposed to hydrometeorological, climatological, and geological hazards.

The *Multihazard Risk Atlas of Maldives* examines the often-unexplored networks of human and environmental paradigms, producing a compound picture of risk that hinders development. It compiles the maps generated for Action on Climate Change in South Asia: Establishing a National Geospatial Database for Mainstreaming Climate Change Adaptation into Development Activities and Policies in Maldives, an Asian Development Bank–Maldives Ministry of Environment technical assistance (TA) project.

The national and atoll-scale maps were produced to paint a clearer picture of climate and disaster risk in Maldives and can hopefully be useful in formulating action plans and policies and in communicating risk information toward averting economic losses and harm. They can become useful tools in visually comparing spatial and temporal variations, spotting interaction among layers, and in guiding possible development directions considering future climate scenarios in the context of the changing geography, demographics, climate, economy, and environmental conditions of Maldives.

This atlas is divided into thematic sections: geography, climate and geophysical hazards, economy and demographics, and biodiversity. A summary volume provides the highlights and key messages from the overall TA activities in Maldives. The main goal of this atlas is to capture often-unmapped factors shaping Maldives' contrasting picture of paradise in the midst of an evolving world, which generates natural hazards to the islands.

Map of the world. A map visualizing the different countries of the world (photo from Crates, World Map without Boundaries, Creative Commons).

Paradise at Risk

In the middle of the Indian Ocean, southwest of India, lies a chain of atolls with islands forming a small low-lying island nation called Maldives. These tropical islands, emerging from corals and sprinkled across turquoise crystal calm waters caressing white sand beaches, paint a picture of a paradise.

The flatlands and surrounding seas form mangrove and aquatic ecosystems supporting the life of the communities that inhabit a tenth of the nation's total islands. These communities have witnessed the continuous formation and transformation of the islands as a product of natural tidal action eroding and depositing sand or of human hands reclaiming or dredging the land. They are also highly vulnerable to strong waves during storms or earthquakes (Waheed and Shakoor 2015). Moreover, a 1-meter high increase in sea level due to a warmer world (Khan et al. 2002) could inundate two-thirds of the country's land (Ahmed and Suphachalasai 2014). Rising seas could wipe away Maldives as most of its islands are less than 3 meters high (Khan et al. 2002). In addition, the natural biota in the islands might not be able to cope with the fast-paced development and could be displaced (ADB 2017). Geologic events and the changing climate pose a threat to the future of the islands and its inhabitants.

This volume examines Maldives, its geography, and the different atolls and their features.

Golden sunset. The usual serene and calm afternoon in a shore of Maldives. This golden reflection of the sun in the crystal clear Maldives' water makes it a good time to sit, relax, and see the beauty of nature (photo by P.K. Niyogi).

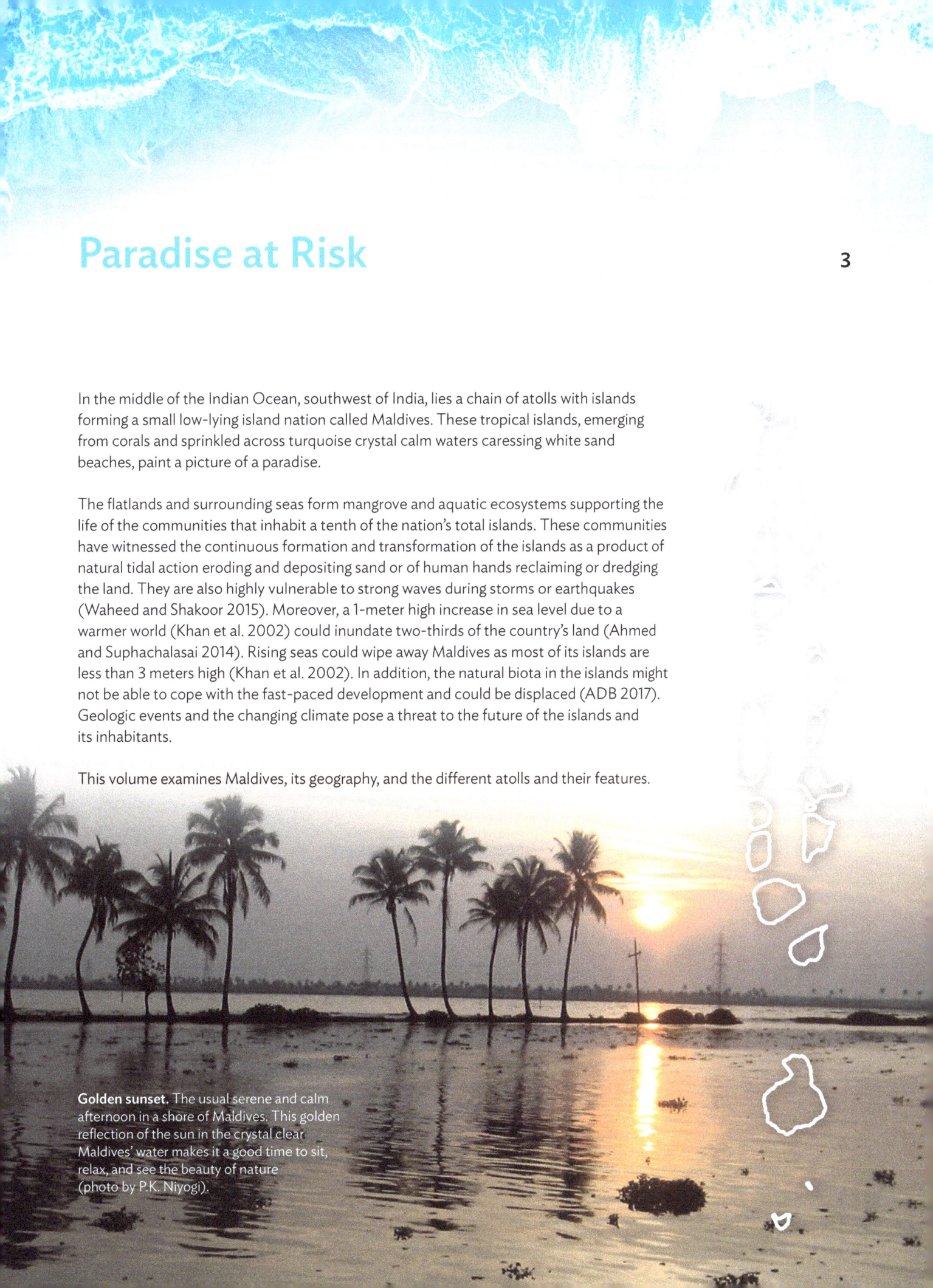

Geography of Maldives

Exclamation Point on the Ocean

Maldives sits like an exclamation point, marking the middle of the Indian Ocean and drawn vertically almost at the equator. Currently, there are about 1,900 islands forming 26 atolls scattered across 870 kilometers from north to south. Less than 200 islands are inhabited. These numbers might change in the coming years due to the dynamic shifts in factors shaping the island such as sea level and land reclamation. Islands rising up to 2.4 meters above mean sea level subtly punctuate the flatness of the sea.

This volume explores the atolls, the transformation of the islands through land reclamation, and how the land is being managed.

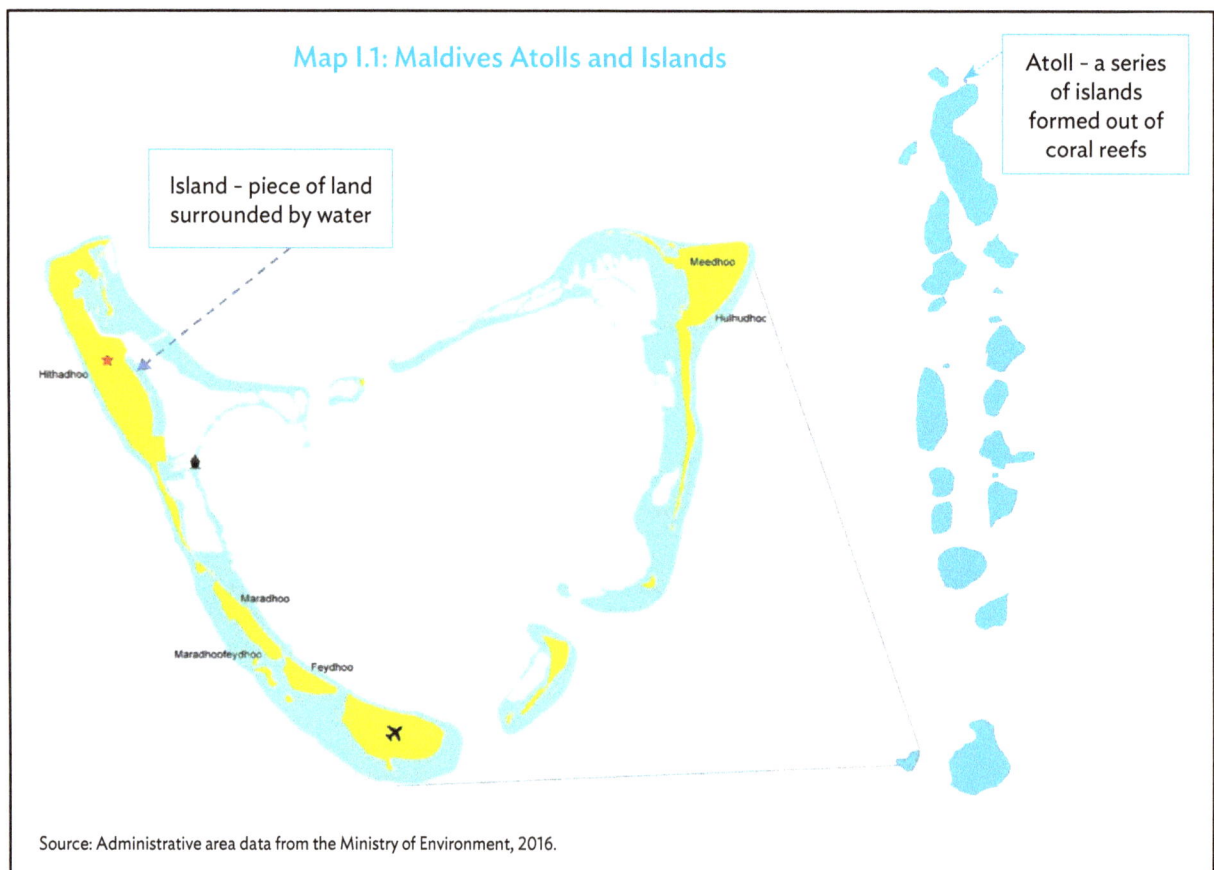

Map I.1: Maldives Atolls and Islands

Atoll - a series of islands formed out of coral reefs

Island - piece of land surrounded by water

Source: Administrative area data from the Ministry of Environment, 2016.

Map I.2: Maldives, Basemap

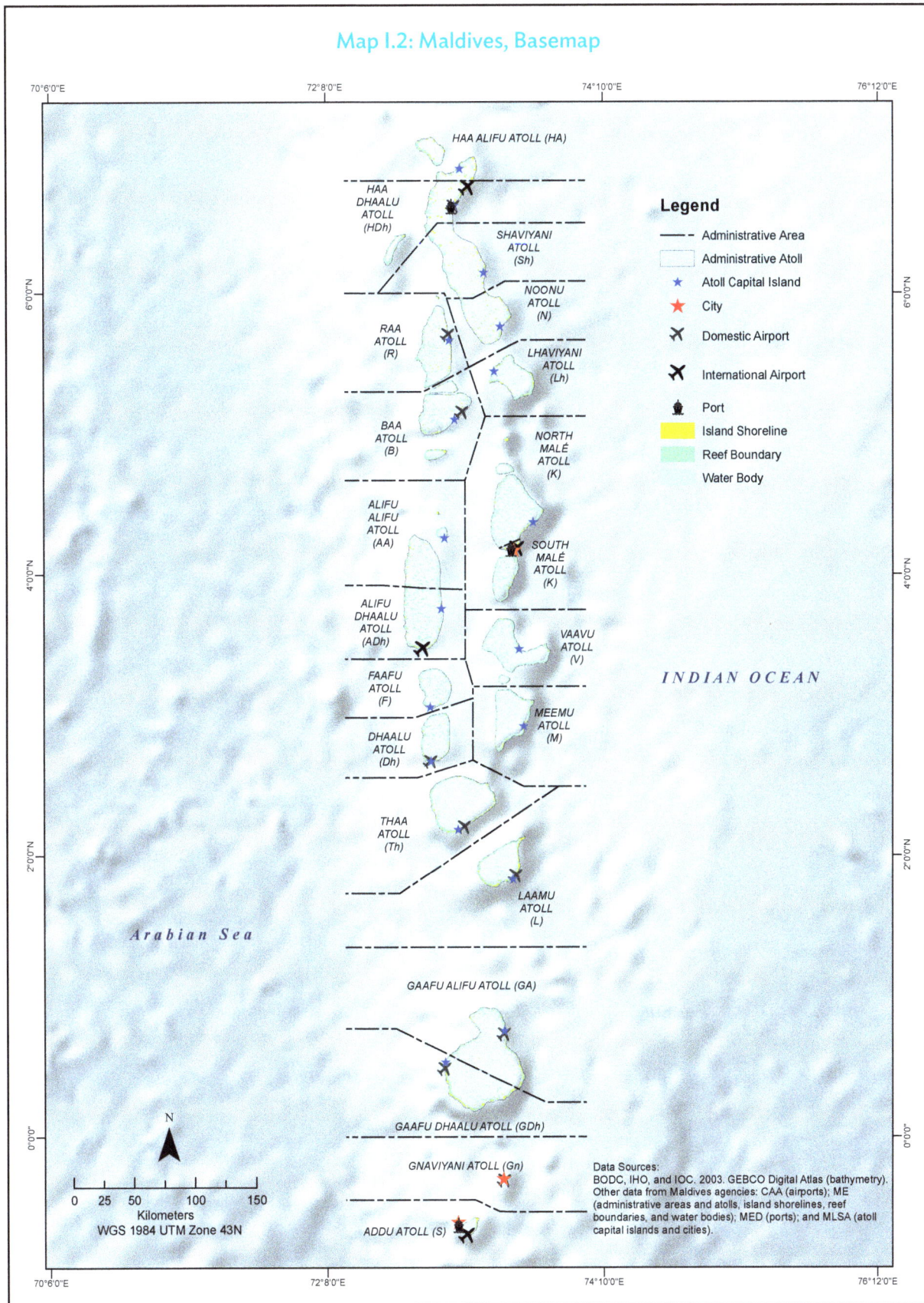

Legend

- Administrative Area
- Administrative Atoll
- ★ Atoll Capital Island
- ★ City
- ✈ Domestic Airport
- ✈ International Airport
- Port
- Island Shoreline
- Reef Boundary
- Water Body

HAA ALIFU ATOLL (HA)
HAA DHAALU ATOLL (HDh)
SHAVIYANI ATOLL (Sh)
NOONU ATOLL (N)
RAA ATOLL (R)
LHAVIYANI ATOLL (Lh)
BAA ATOLL (B)
NORTH MALÉ ATOLL (K)
ALIFU ALIFU ATOLL (AA)
SOUTH MALÉ ATOLL (K)
ALIFU DHAALU ATOLL (ADh)
VAAVU ATOLL (V)
FAAFU ATOLL (F)
MEEMU ATOLL (M)
DHAALU ATOLL (Dh)
THAA ATOLL (Th)
LAAMU ATOLL (L)
GAAFU ALIFU ATOLL (GA)
GAAFU DHAALU ATOLL (GDh)
GNAVIYANI ATOLL (Gn)
ADDU ATOLL (S)

INDIAN OCEAN
Arabian Sea

N

0 25 50 100 150
Kilometers
WGS 1984 UTM Zone 43N

Data Sources:
BODC, IHO, and IOC. 2003. GEBCO Digital Atlas (bathymetry). Other data from Maldives agencies: CAA (airports); ME (administrative areas and atolls, island shorelines, reef boundaries, and water bodies); MED (ports); and MLSA (atoll capital islands and cities).

Map I.3: Addu City, Basemap

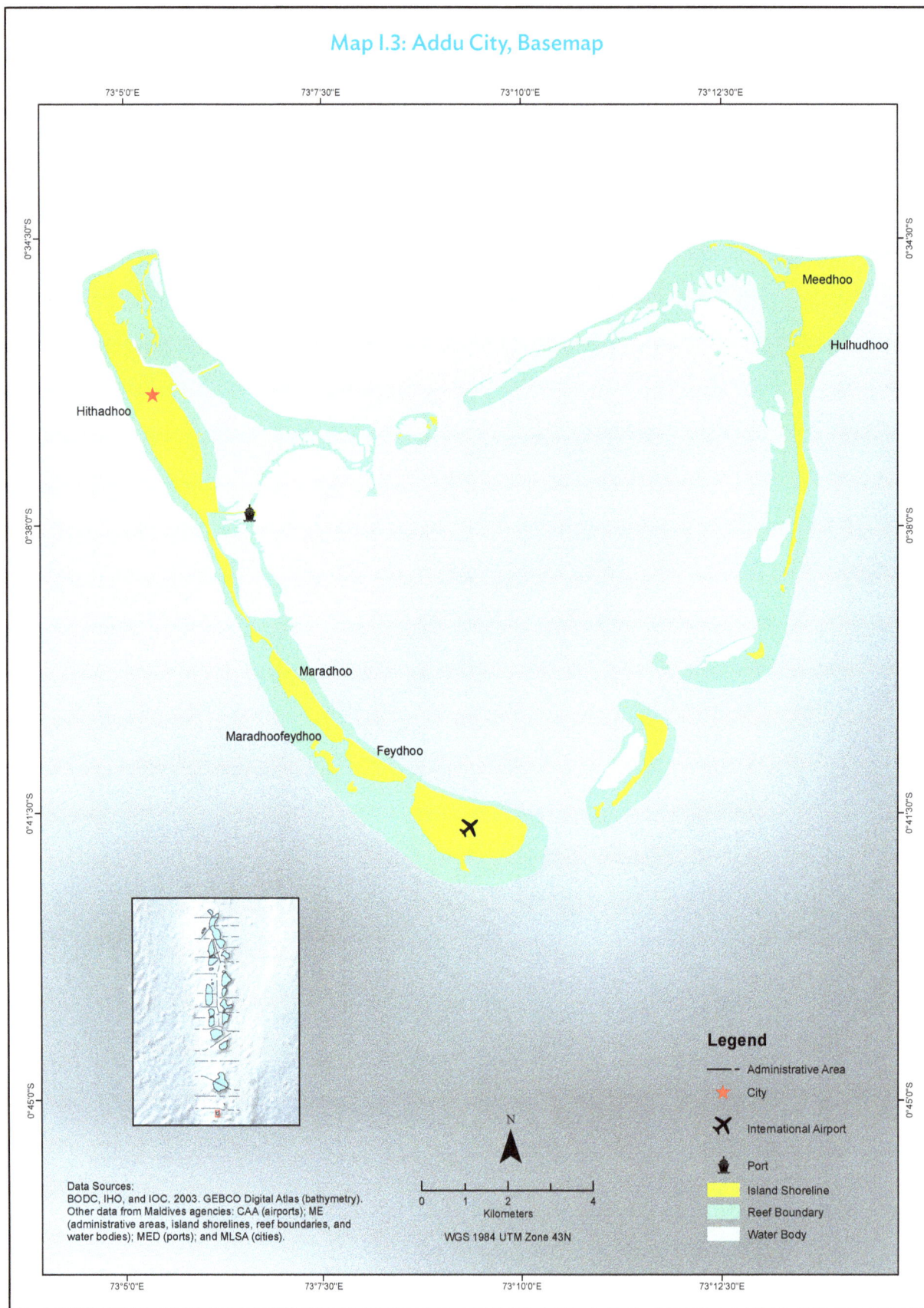

Meedhoo

Hulhudhoo

Hithadhoo

Maradhoo

Maradhoofeydhoo

Feydhoo

Data Sources:
BODC, IHO, and IOC. 2003. GEBCO Digital Atlas (bathymetry).
Other data from Maldives agencies: CAA (airports); ME
(administrative areas, island shorelines, reef boundaries, and
water bodies); MED (ports); and MLSA (cities).

N

0   1   2   4
Kilometers

WGS 1984 UTM Zone 43N

Legend

- - - - Administrative Area

★   City

✈   International Airport

⚓   Port

　 Island Shoreline

　 Reef Boundary

　 Water Body

Map I.4: Alifu Alifu Atoll, Basemap

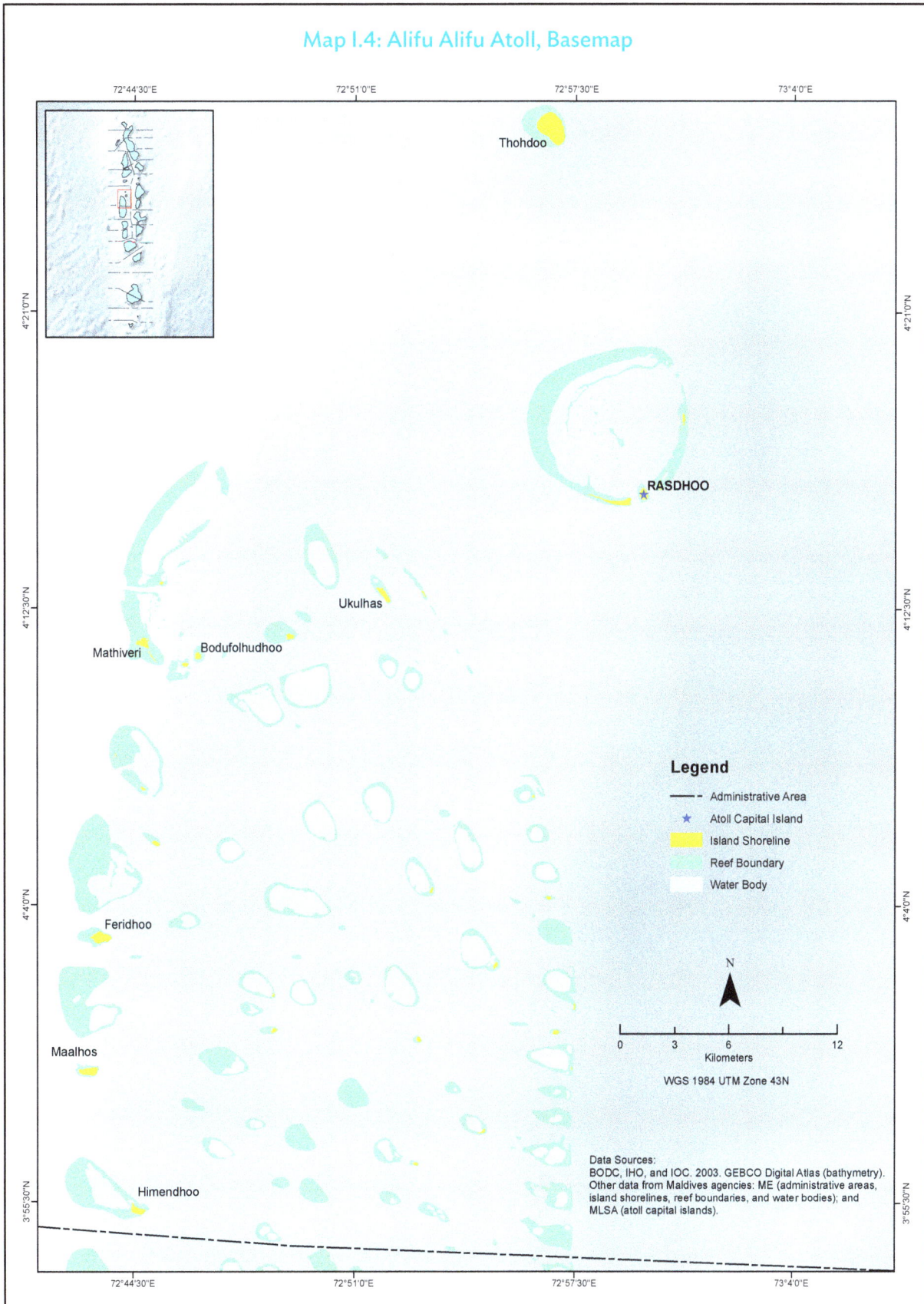

Thohdoo

RASDHOO

Ukulhas

Mathiveri

Bodufolhudhoo

Feridhoo

Maalhos

Himendhoo

Legend

— — — Administrative Area

★ Atoll Capital Island

Island Shoreline

Reef Boundary

Water Body

N

0   3   6   12
Kilometers

WGS 1984 UTM Zone 43N

Data Sources:
BODC, IHO, and IOC. 2003. GEBCO Digital Atlas (bathymetry).
Other data from Maldives agencies: ME (administrative areas,
island shorelines, reef boundaries, and water bodies); and
MLSA (atoll capital islands).

Map I.5: Alifu Dhaalu Atoll, Basemap

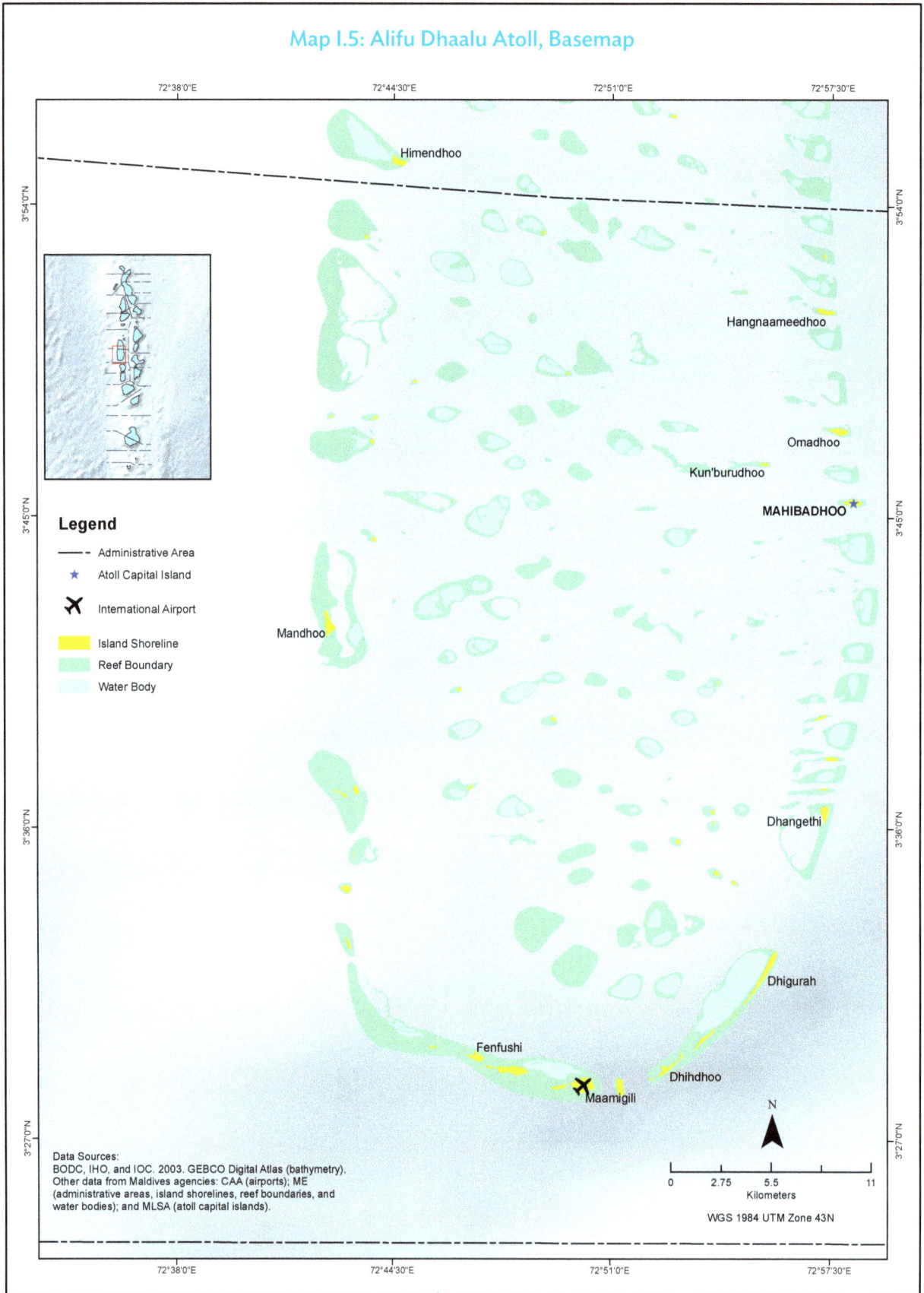

72°38'0"E   72°44'30"E   72°51'0"E   72°57'30"E

Himendhoo

Hangnaameedhoo

Omadhoo

Kun'burudhoo

MAHIBADHOO ★

Mandhoo

3°45'0"N

3°54'0"N

Legend

— · — Administrative Area

★ Atoll Capital Island

✈ International Airport

▮ Island Shoreline

▮ Reef Boundary

▮ Water Body

Dhangethi

3°36'0"N

Dhigurah

Fenfushi

Dhihdhoo

✈ Maamigili

N

3°27'0"N

Data Sources:
BODC, IHO, and IOC. 2003. GEBCO Digital Atlas (bathymetry).
Other data from Maldives agencies: CAA (airports); ME
(administrative areas, island shorelines, reef boundaries, and
water bodies); and MLSA (atoll capital islands).

0   2.75   5.5   11
Kilometers

WGS 1984 UTM Zone 43N

72°38'0"E   72°44'30"E   72°51'0"E   72°57'30"E

Map I.6: Baa Atoll, Basemap

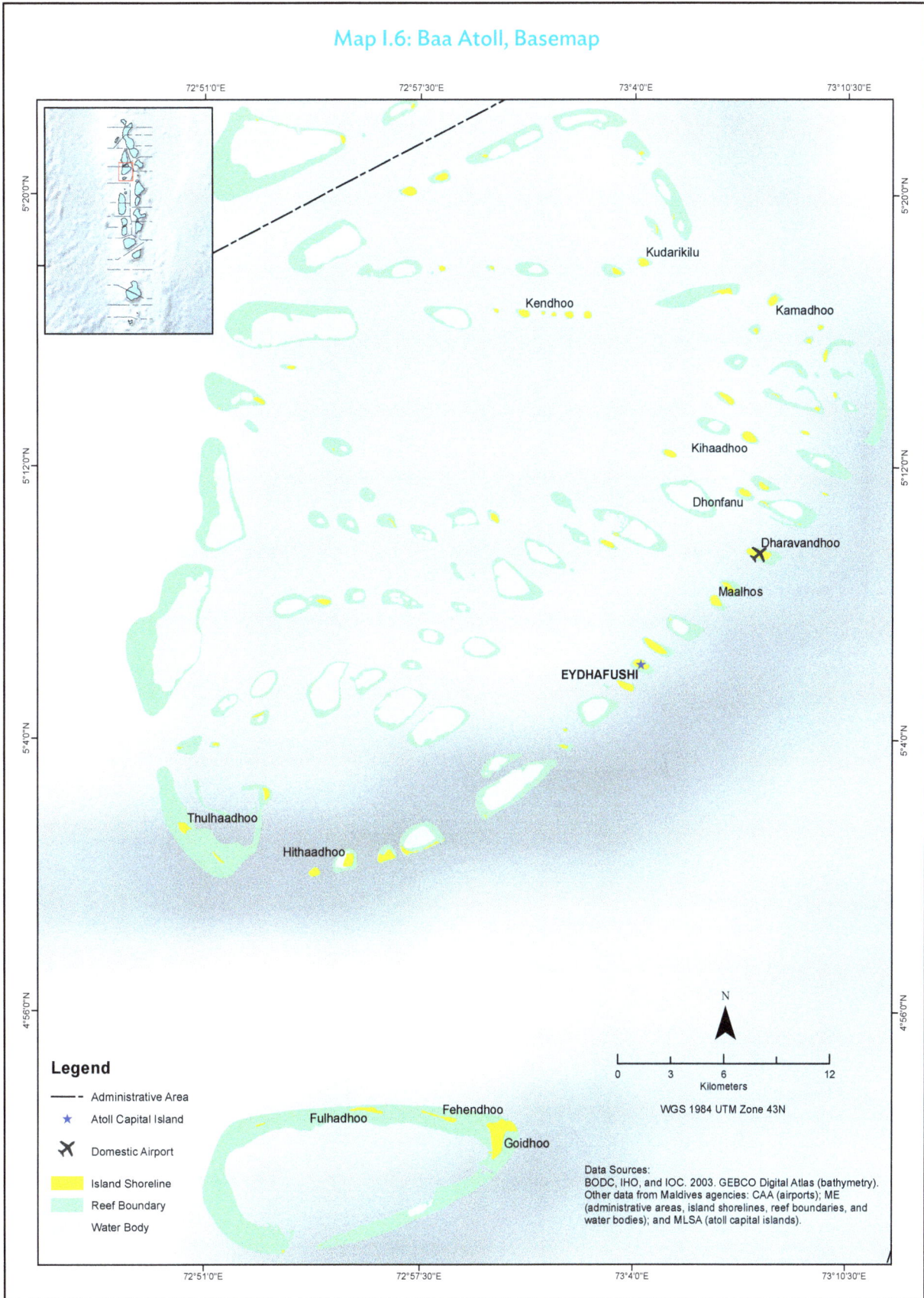

Kudarikilu

Kendhoo

Kamadhoo

Kihaadhoo

Dhonfanu

Dharavandhoo

Maalhos

EYDHAFUSHI

Thulhaadhoo

Hithaadhoo

N

0   3   6   12
Kilometers

WGS 1984 UTM Zone 43N

Fulhadhoo   Fehendhoo

Goidhoo

Legend

- - - - Administrative Area

★   Atoll Capital Island

✈   Domestic Airport

▮   Island Shoreline

▮   Reef Boundary

Water Body

Data Sources:
BODC, IHO, and IOC. 2003. GEBCO Digital Atlas (bathymetry).
Other data from Maldives agencies: CAA (airports); ME
(administrative areas, island shorelines, reef boundaries, and
water bodies); and MLSA (atoll capital islands).

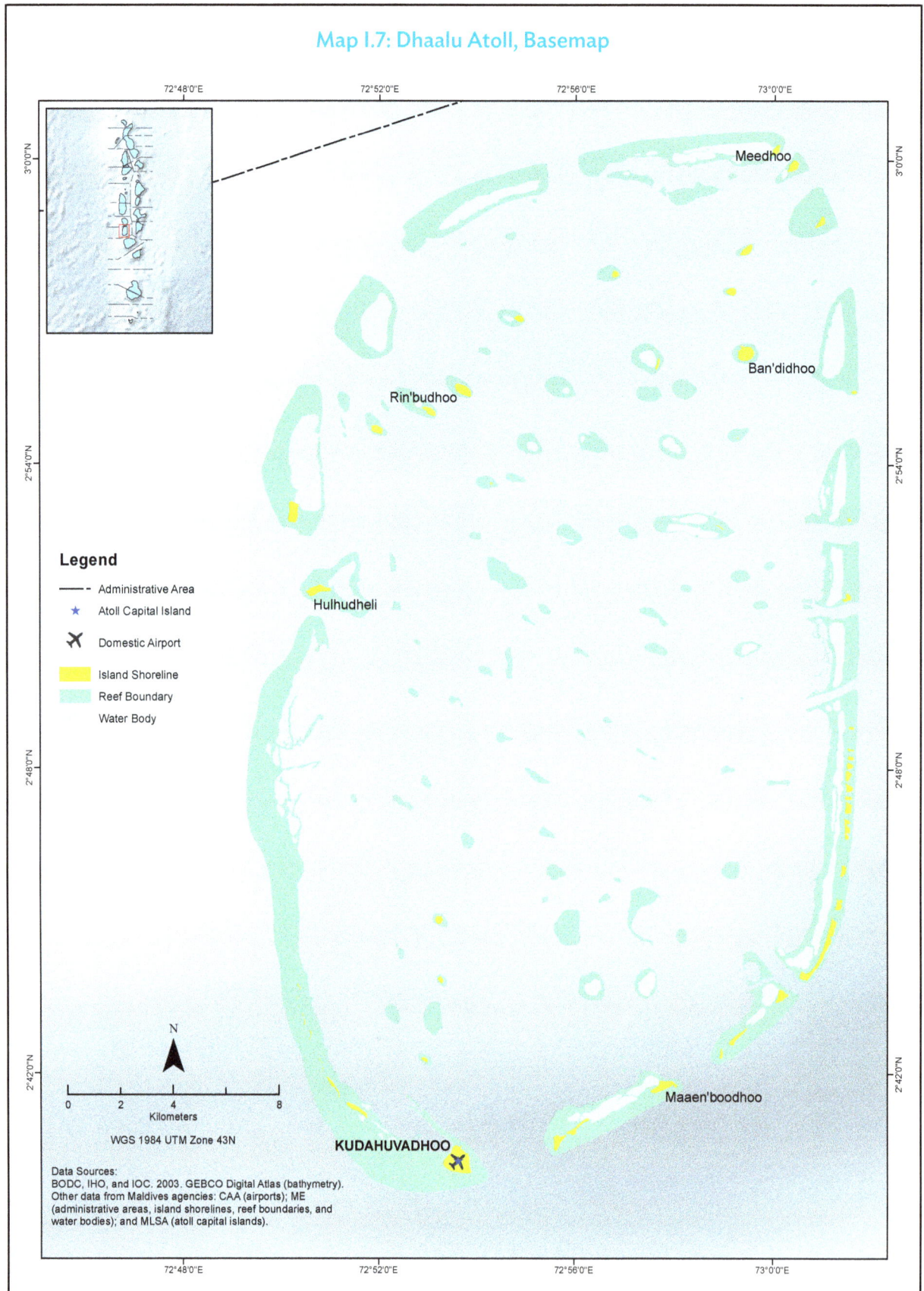

Map I.7: Dhaalu Atoll, Basemap

Meedhoo

Ban'didhoo

Rin'budhoo

Hulhudheli

Legend

— - — Administrative Area

★ Atoll Capital Island

✈ Domestic Airport

■ Island Shoreline

■ Reef Boundary

Water Body

Maaen'boodhoo

N

0   2   4   8
Kilometers

WGS 1984 UTM Zone 43N

KUDAHUVADHOO ✈

Data Sources:
BODC, IHO, and IOC. 2003. GEBCO Digital Atlas (bathymetry).
Other data from Maldives agencies: CAA (airports); ME
(administrative areas, island shorelines, reef boundaries, and
water bodies); and MLSA (atoll capital islands).

Map I.8: Faafu Atoll, Basemap

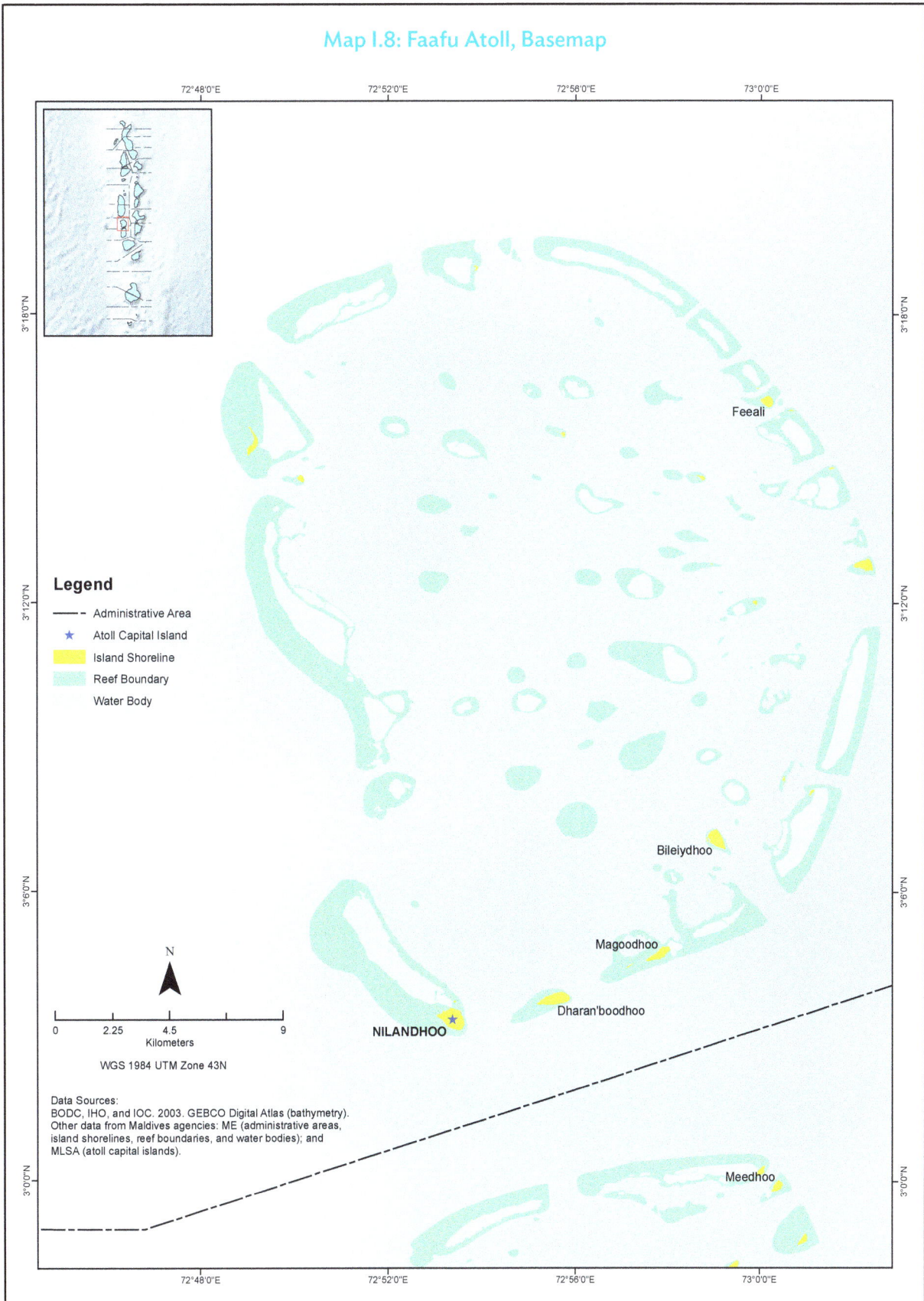

Legend
- – – Administrative Area
- ★ Atoll Capital Island
- Island Shoreline
- Reef Boundary
- Water Body

Feeali

Bileiydhoo

Magoodhoo

Dharan'boodhoo

NILANDHOO

Meedhoo

0   2.25   4.5   9
Kilometers

WGS 1984 UTM Zone 43N

Data Sources:
BODC, IHO, and IOC. 2003. GEBCO Digital Atlas (bathymetry).
Other data from Maldives agencies: ME (administrative areas,
island shorelines, reef boundaries, and water bodies); and
MLSA (atoll capital islands).

Map I.9: Gaafu Alifu Atoll, Basemap

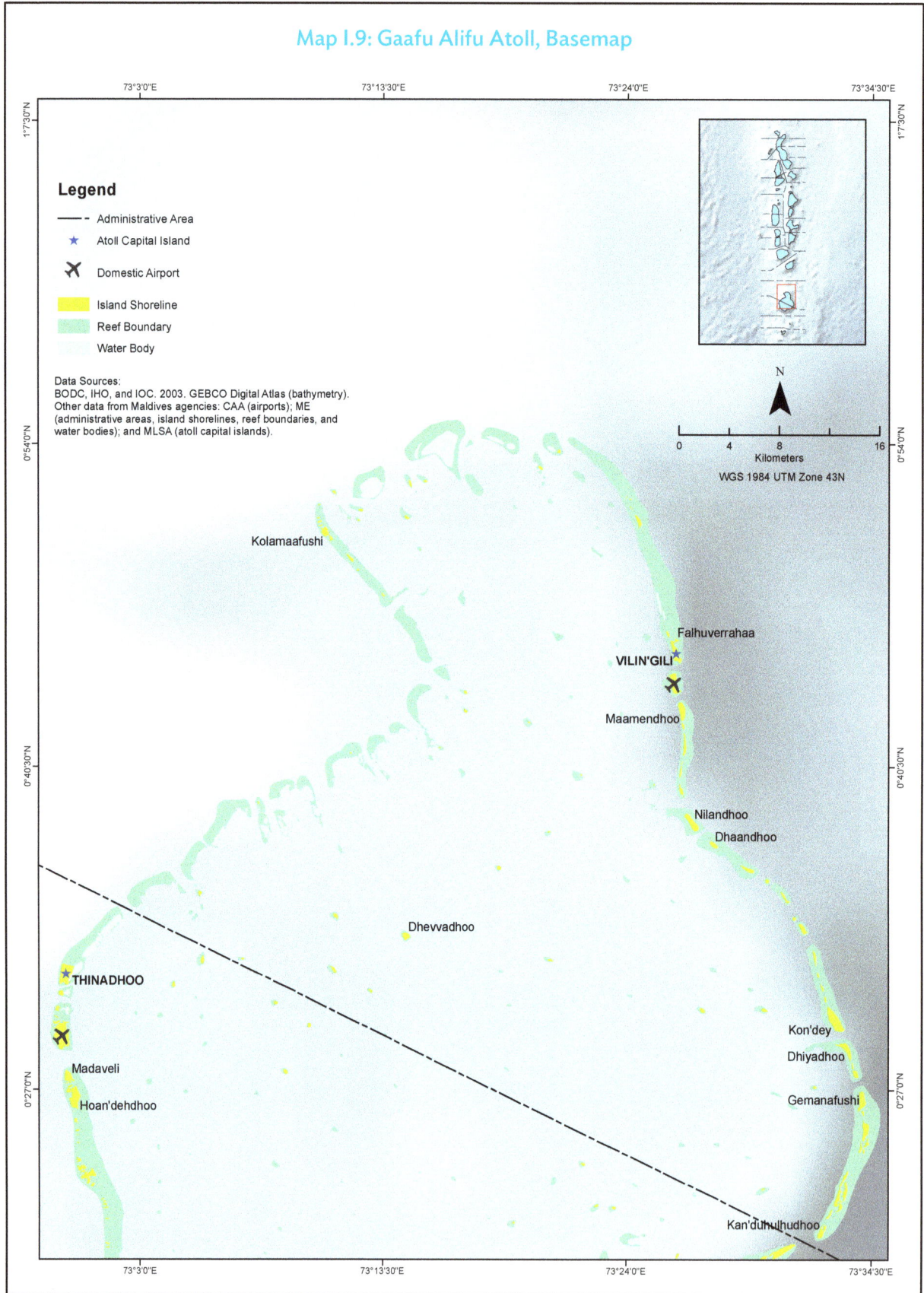

Legend

- – – Administrative Area
- ★ Atoll Capital Island
- ✈ Domestic Airport
- ▨ Island Shoreline
- ▨ Reef Boundary
- ▨ Water Body

Data Sources:
BODC, IHO, and IOC. 2003. GEBCO Digital Atlas (bathymetry).
Other data from Maldives agencies: CAA (airports); ME
(administrative areas, island shorelines, reef boundaries, and
water bodies); and MLSA (atoll capital islands).

N

0   4   8   16
Kilometers
WGS 1984 UTM Zone 43N

Kolamaafushi

Falhuverrahaa
VILIN'GILI ★ ✈

Maamendhoo

Nilandhoo
Dhaandhoo

Dhevvadhoo

★ THINADHOO

✈

Madaveli

Kon'dey

Dhiyadhoo

Hoan'dehdhoo

Gemanafushi

Kan'duhulhudhoo

Map I.10: Gaafu Dhaalu Atoll, Basemap

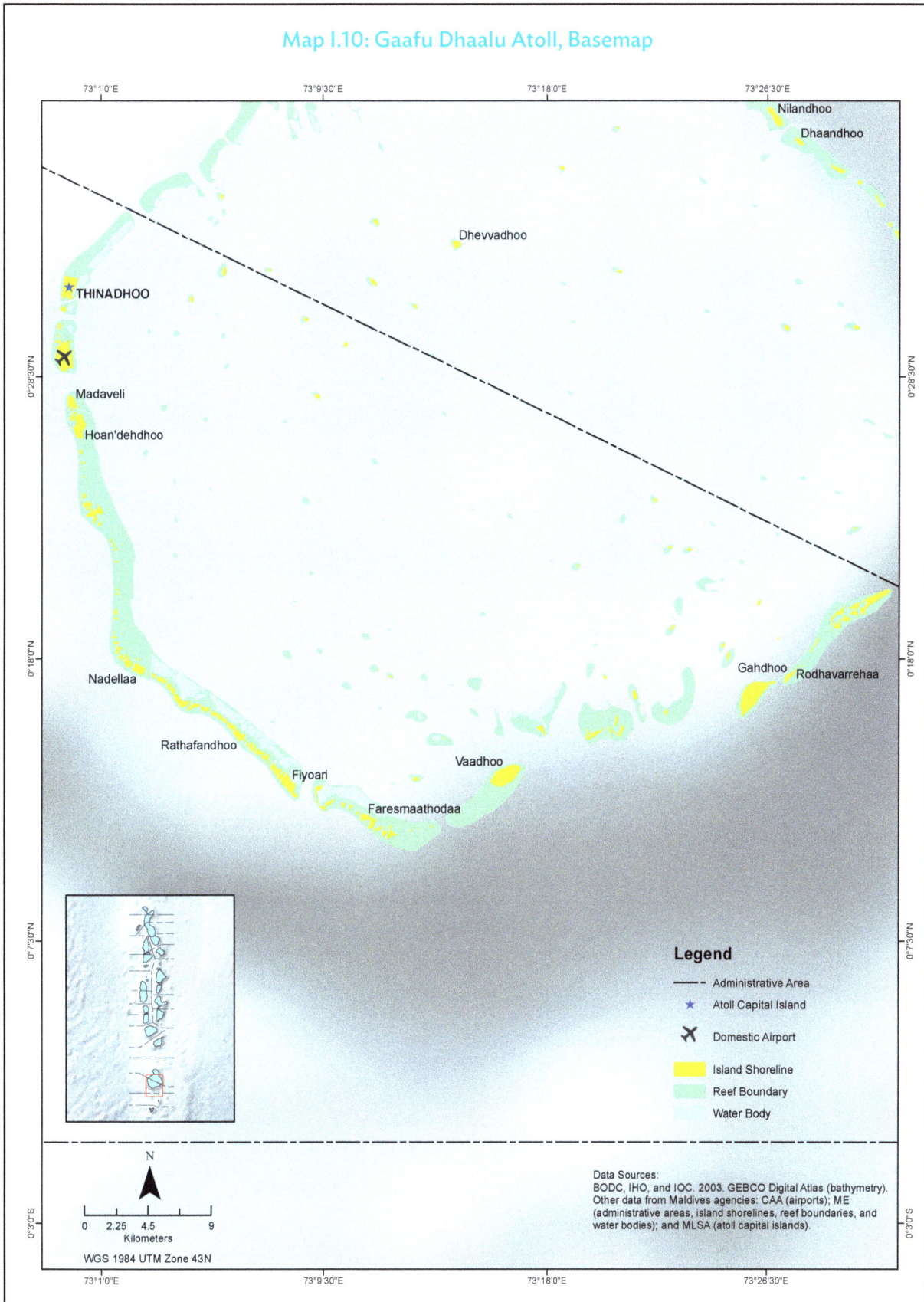

73°1'0"E   73°9'30"E   73°18'0"E   73°26'30"E

Nilandhoo

Dhaandhoo

Dhevvadhoo

★ THINADHOO

0°28'30"N

✈

Madaveli

Hoan'dehdhoo

0°28'30"N

Nadellaa

Gahdhoo Rodhavarrehaa

0°18'0"N

0°18'0"N

Rathafandhoo

Fiyoari

Vaadhoo

Faresmaathodaa

0°7'30"N

0°7'30"N

Legend

- - - Administrative Area

★ Atoll Capital Island

✈ Domestic Airport

▮ Island Shoreline

▮ Reef Boundary

Water Body

N

0 2.25 4.5   9
Kilometers

WGS 1984 UTM Zone 43N

0°3'0"S

0°3'0"S

Data Sources:
BODC, IHO, and IOC. 2003. GEBCO Digital Atlas (bathymetry).
Other data from Maldives agencies: CAA (airports); ME
(administrative areas, island shorelines, reef boundaries, and
water bodies); and MLSA (atoll capital islands).

73°1'0"E   73°9'30"E   73°18'0"E   73°26'30"E

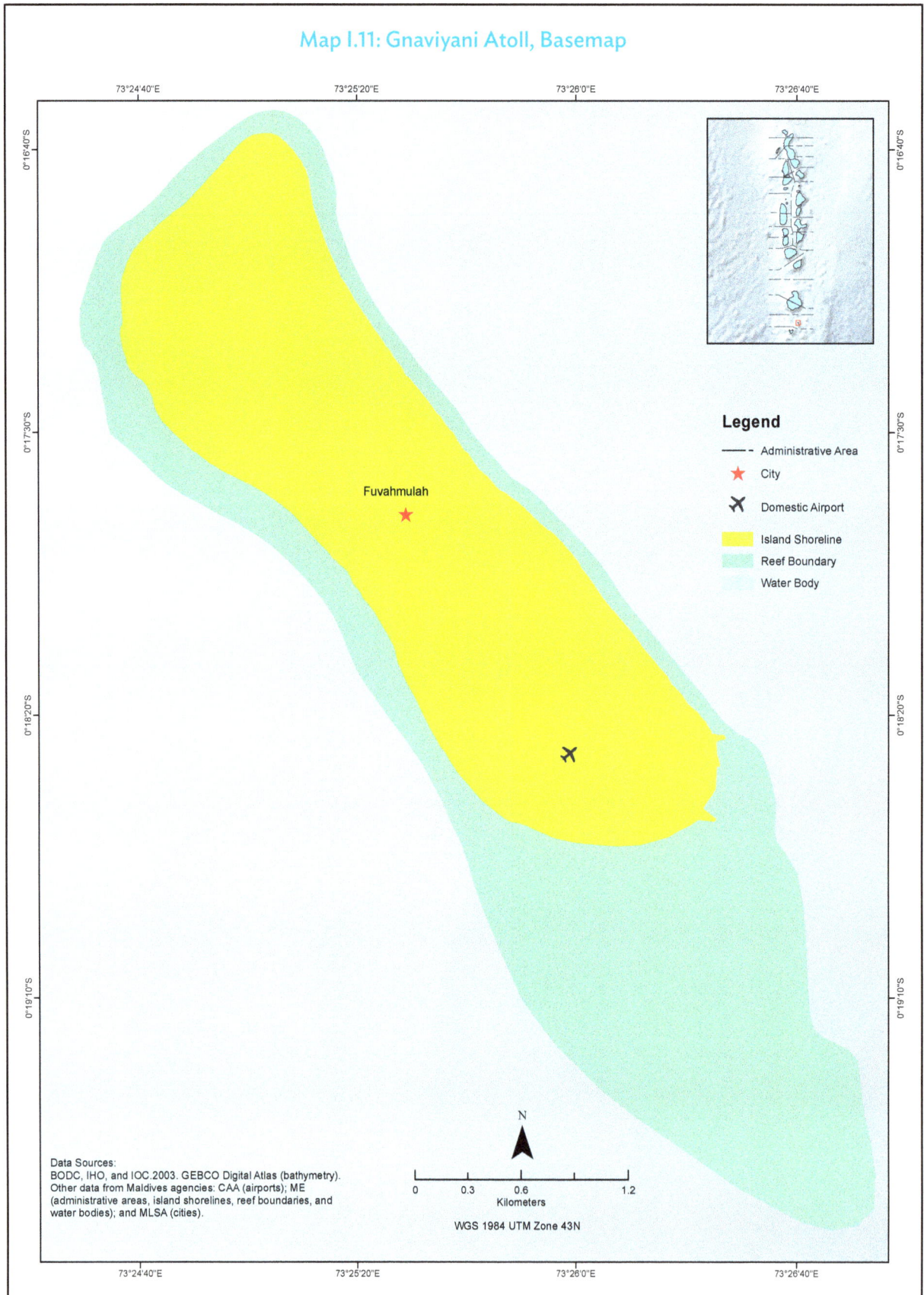

Map I.11: Gnaviyani Atoll, Basemap

73°24'40"E   73°25'20"E   73°26'0"E   73°26'40"E

0°16'40"S

0°17'30"S

Fuvahmulah
★

0°18'20"S

✈

0°19'10"S

Legend

— · — Administrative Area

★ City

✈ Domestic Airport

▮ Island Shoreline

▮ Reef Boundary

▮ Water Body

Data Sources:
BODC, IHO, and IOC.2003. GEBCO Digital Atlas (bathymetry).
Other data from Maldives agencies: CAA (airports); ME
(administrative areas, island shorelines, reef boundaries, and
water bodies); and MLSA (cities).

N

0   0.3   0.6   1.2
Kilometers

WGS 1984 UTM Zone 43N

73°24'40"E   73°25'20"E   73°26'0"E   73°26'40"E

Map I.12: Haa Alifu Atoll, Basemap

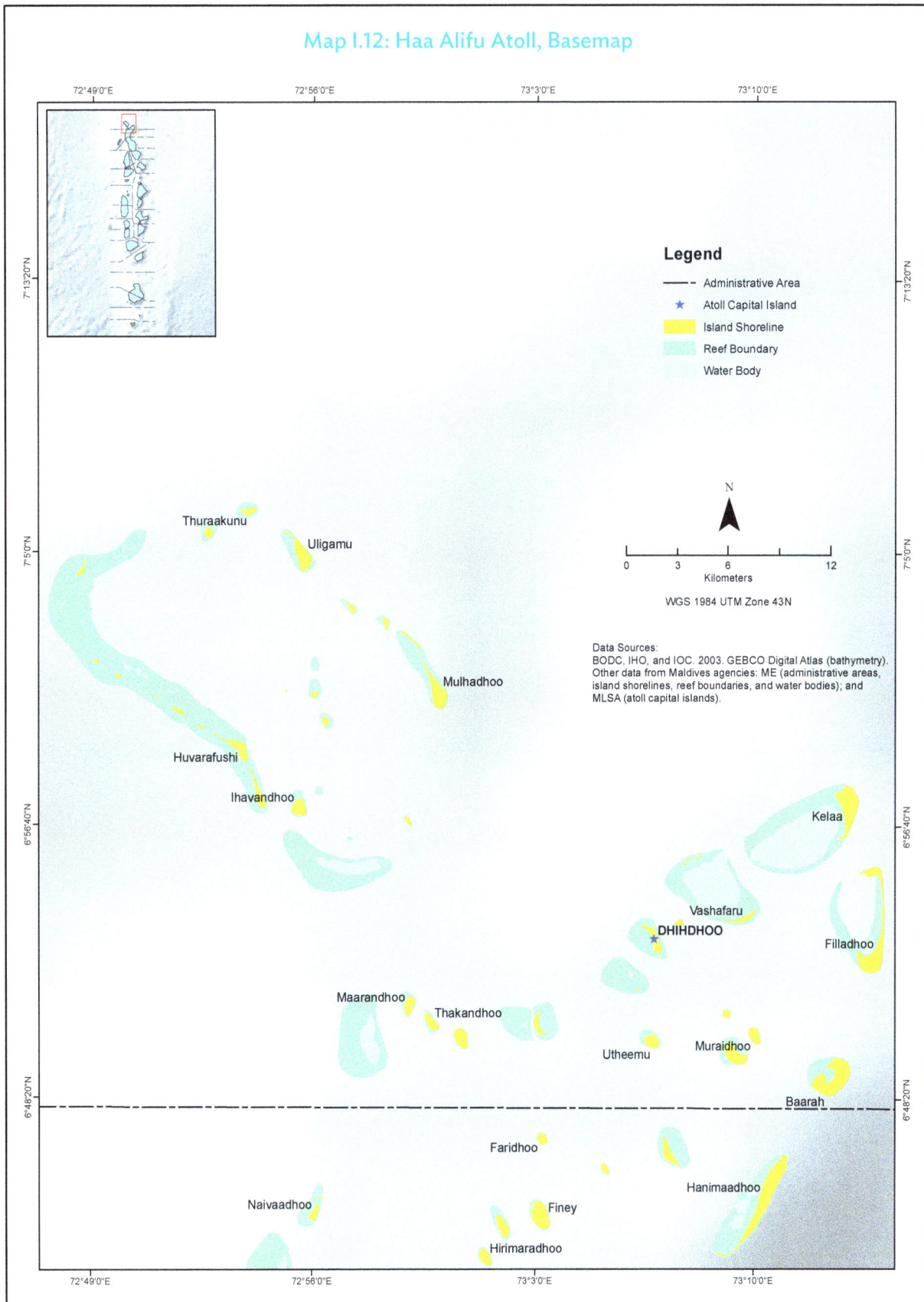

72°49'0"E   72°56'0"E   73°3'0"E   73°10'0"E

Legend

- – – Administrative Area
- ★ Atoll Capital Island
- Island Shoreline
- Reef Boundary
- Water Body

N

0   3   6   12
Kilometers

WGS 1984 UTM Zone 43N

Data Sources:
BODC, IHO, and IOC. 2003. GEBCO Digital Atlas (bathymetry).
Other data from Maldives agencies: ME (administrative areas,
island shorelines, reef boundaries, and water bodies); and
MLSA (atoll capital islands).

Thuraakunu

Uligamu

Mulhadhoo

Huvarafushi

Ihavandhoo

Kelaa

Vashafaru

DHIHDHOO ★

Filladhoo

Maarandhoo

Thakandhoo

Utheemu

Muraidhoo

Baarah

Faridhoo

Hanimaadhoo

Naivaadhoo

Finey

Hirimaradhoo

7°13'20"N   7°5'0"N   6°56'40"N   6°48'20"N

Map I.13: Haa Dhaalu Atoll, Basemap

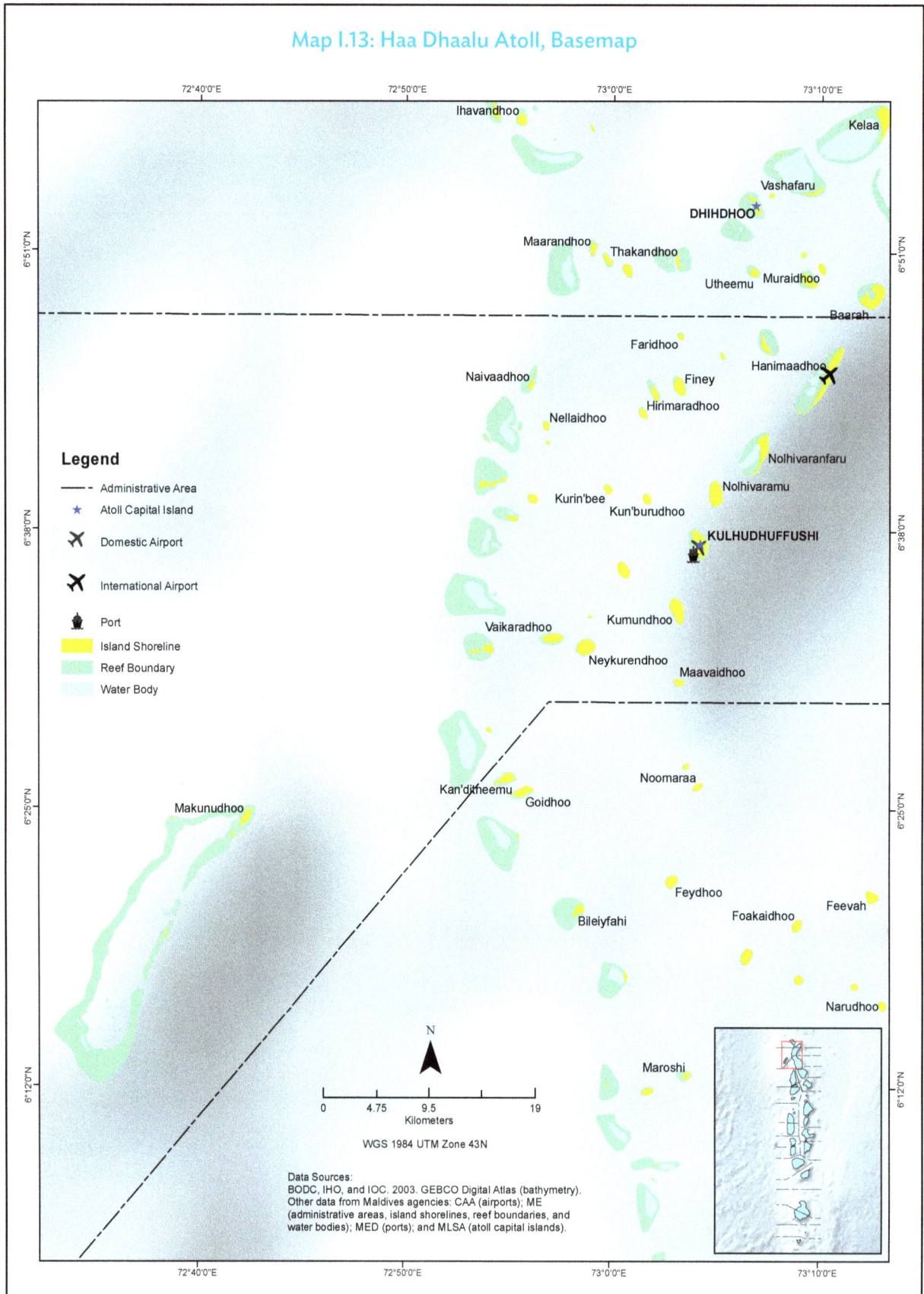

Ihavandhoo

Kelaa

Vashafaru

DHIHDHOO ★

Maarandhoo

Thakandhoo

Utheemu   Muraidhoo

Baarah

Faridhoo

Naivaadhoo

Hanimaadhoo ✈

Finey

Hirimaradhoo

Nellaidhoo

Nolhivaranfaru

Nolhivaramu

Kurin'bee

Kun'burudhoo

KULHUDHUFFUSHI ✈⚓

Kumundhoo

Vaikaradhoo

Neykurendhoo

Maavaidhoo

Noomaraa

Kan'ditheemu

Makunudhoo

Goidhoo

Feydhoo

Feevah

Bileiyfahi

Foakaidhoo

Narudhoo

Maroshi

Legend

– – – Administrative Area
★ Atoll Capital Island
✈ Domestic Airport
✈ International Airport
⚓ Port
▮ Island Shoreline
▮ Reef Boundary
▮ Water Body

N
▲

```
0      4.75      9.5              19
            Kilometers
```

WGS 1984 UTM Zone 43N

Data Sources:
BODC, IHO, and IOC. 2003. GEBCO Digital Atlas (bathymetry).
Other data from Maldives agencies: CAA (airports); ME
(administrative areas, island shorelines, reef boundaries, and
water bodies); MED (ports); and MLSA (atoll capital islands).

Map I.14: Laamu Atoll, Basemap

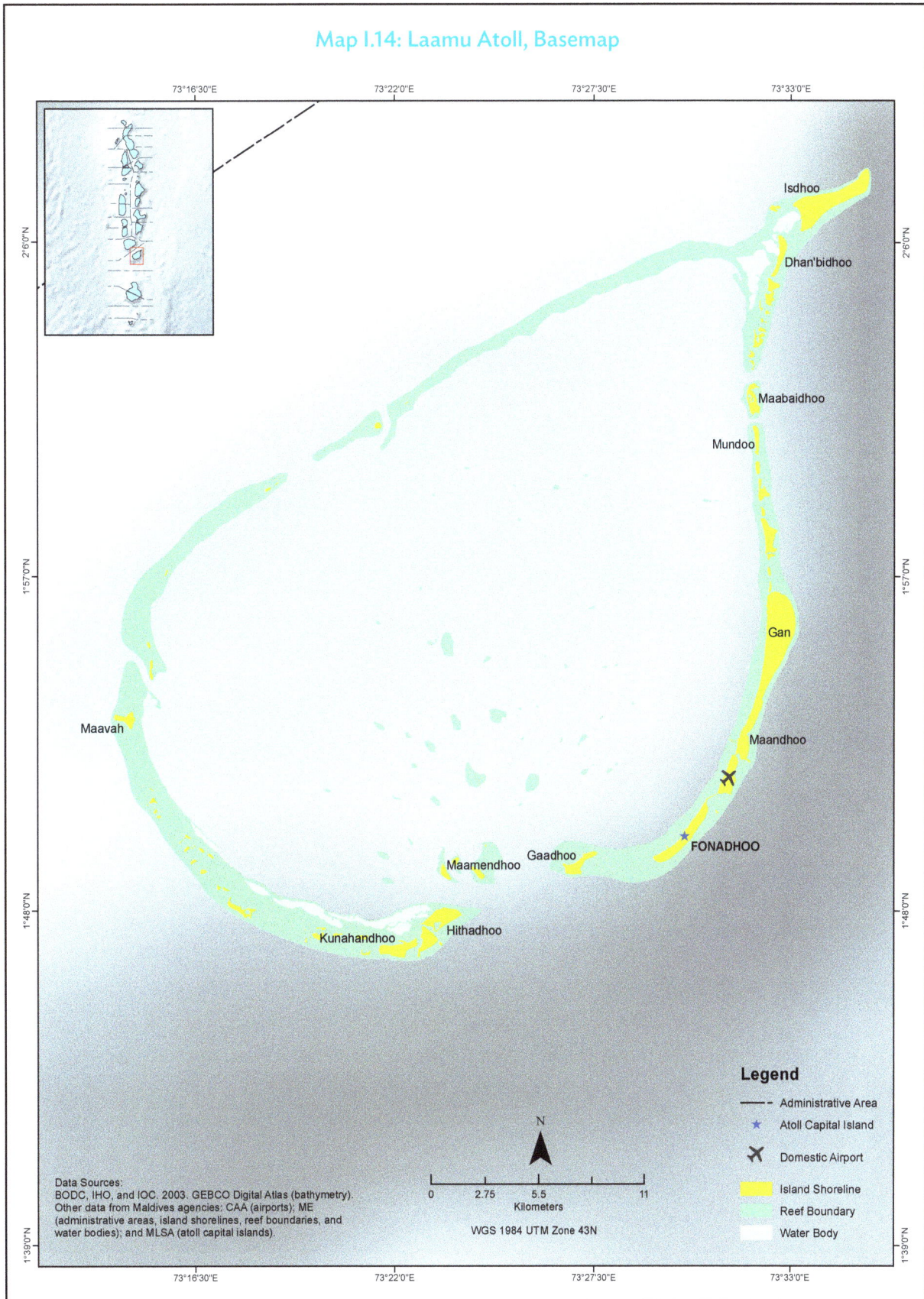

73°16'30"E   73°22'0"E   73°27'30"E   73°33'0"E

2°6'0"N

1°57'0"N

1°48'0"N

1°39'0"N

Isdhoo

Dhan'bidhoo

Maabaidhoo

Mundoo

Gan

Maandhoo

Maavah

FONADHOO

Gaadhoo

Maamendhoo

Kunahandhoo   Hithadhoo

Legend

- — · — Administrative Area
- ★ Atoll Capital Island
- ✕ Domestic Airport
- Island Shoreline
- Reef Boundary
- Water Body

N

0   2.75   5.5   11
Kilometers

WGS 1984 UTM Zone 43N

Data Sources:
BODC, IHO, and IOC. 2003. GEBCO Digital Atlas (bathymetry).
Other data from Maldives agencies: CAA (airports); ME
(administrative areas, island shorelines, reef boundaries, and
water bodies); and MLSA (atoll capital islands).

Map I.15: Lhaviyani Atoll, Basemap

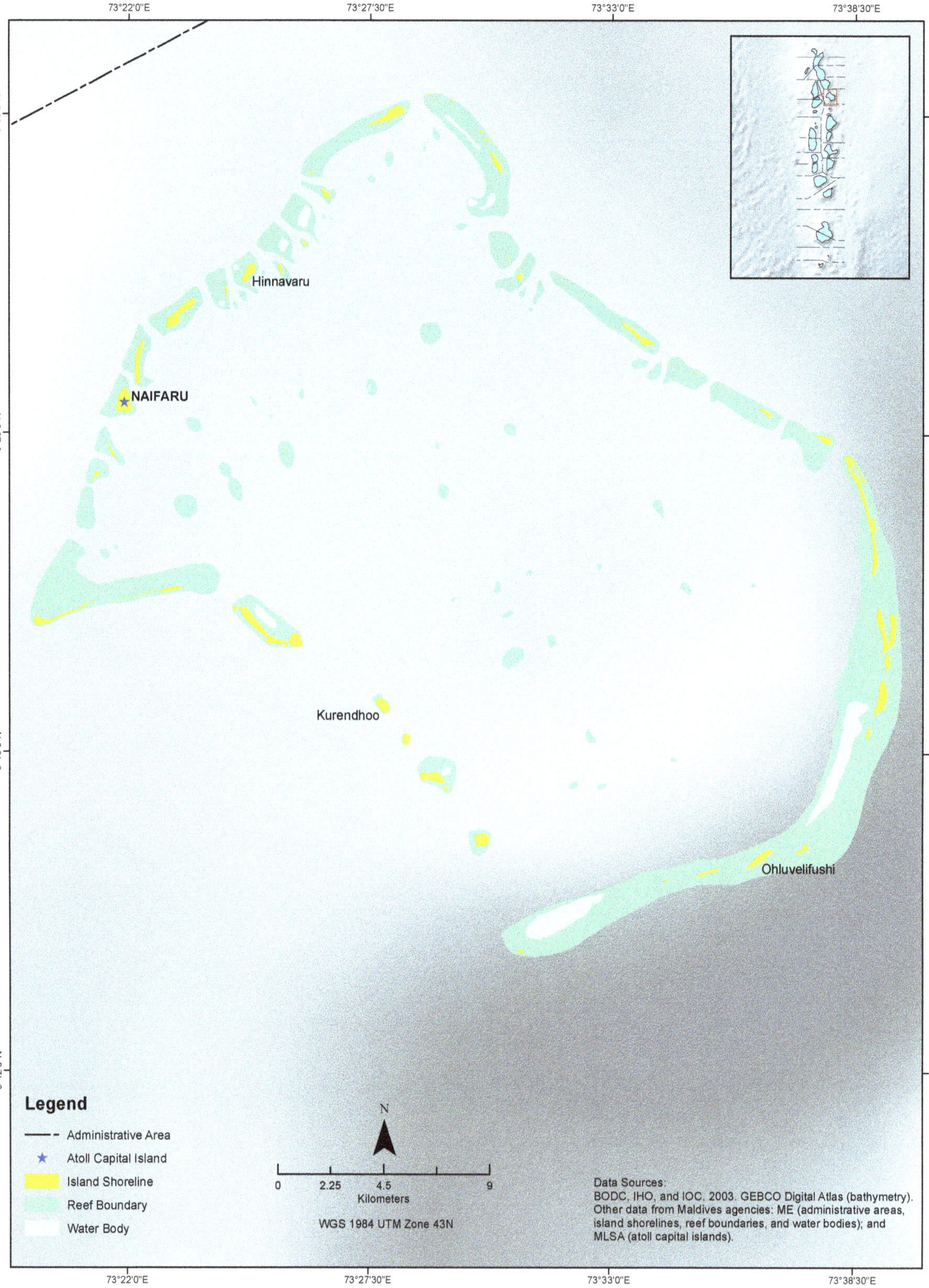
73°22'0"E
73°27'30"E
73°33'0"E
73°38'30"E
5°33'0"N
5°26'0"N
5°19'0"N
5°12'0"N
Hinnavaru
NAIFARU
Kurendhoo
Ohluvelifushi
Legend
Administrative Area
Atoll Capital Island
Island Shoreline
Reef Boundary
Water Body
N
0
2.25
4.5
9
Kilometers
WGS 1984 UTM Zone 43N
Data Sources:
BODC, IHO, and IOC. 2003. GEBCO Digital Atlas (bathymetry).
Other data from Maldives agencies: ME (administrative areas, island shorelines, reef boundaries, and water bodies); and MLSA (atoll capital islands).

Map I.16: Meemu Atoll, Basemap

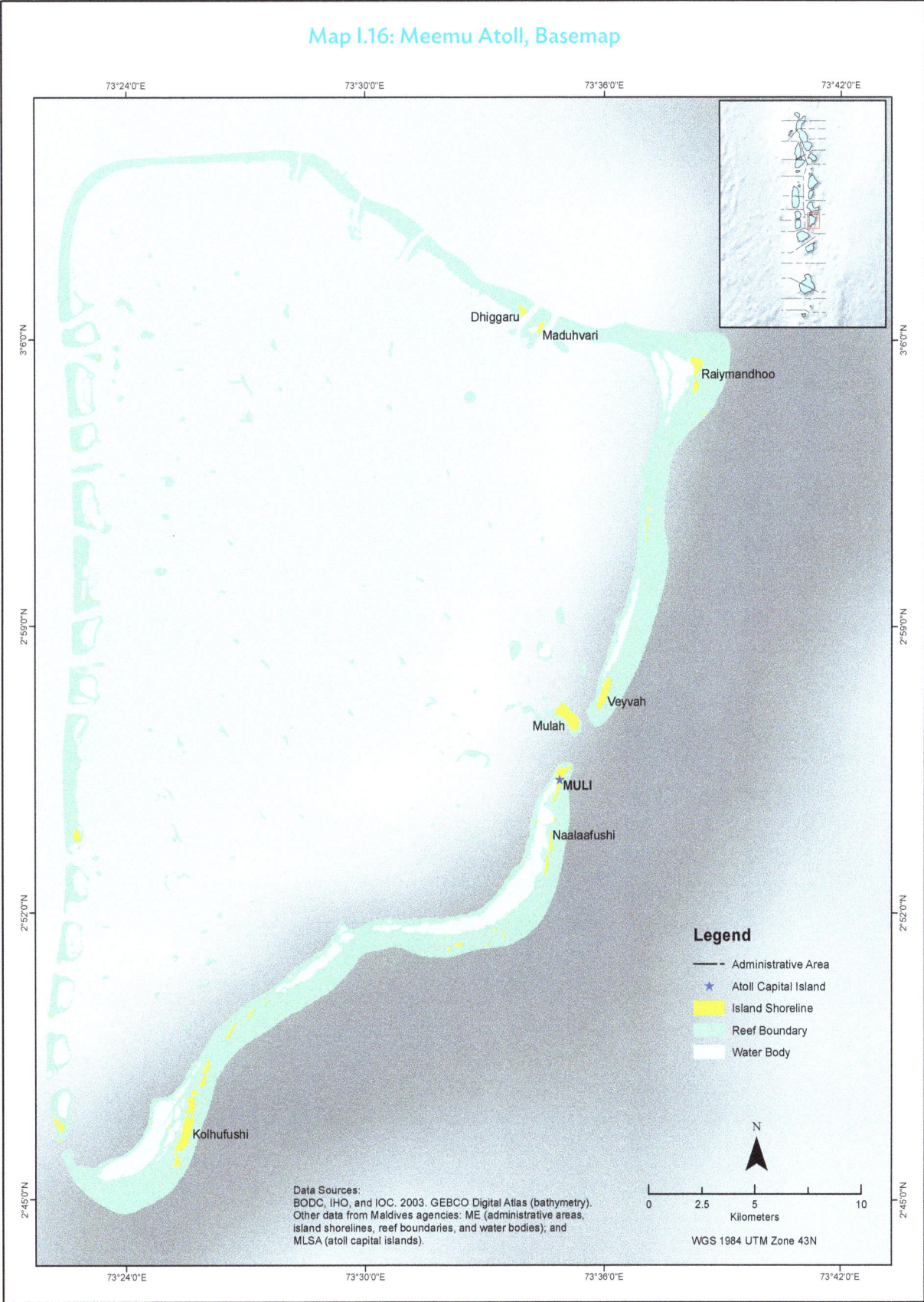

73°24'0"E   73°30'0"E   73°36'0"E   73°42'0"E

3°6'0"N

Dhiggaru

Maduhvari

Raiymandhoo

2°59'0"N

Veyvah

Mulah

★MULI

Naalaafushi

2°52'0"N

Kolhufushi

Legend

— - Administrative Area

★ Atoll Capital Island

▮ Island Shoreline

▮ Reef Boundary

▮ Water Body

N

Data Sources:
BODC, IHO, and IOC. 2003. GEBCO Digital Atlas (bathymetry).
Other data from Maldives agencies: ME (administrative areas,
island shorelines, reef boundaries, and water bodies); and
MLSA (atoll capital islands).

0   2.5   5   10
            Kilometers

WGS 1984 UTM Zone 43N

73°24'0"E   73°30'0"E   73°36'0"E   73°42'0"E

Map I.17: Noonu Atoll, Basemap

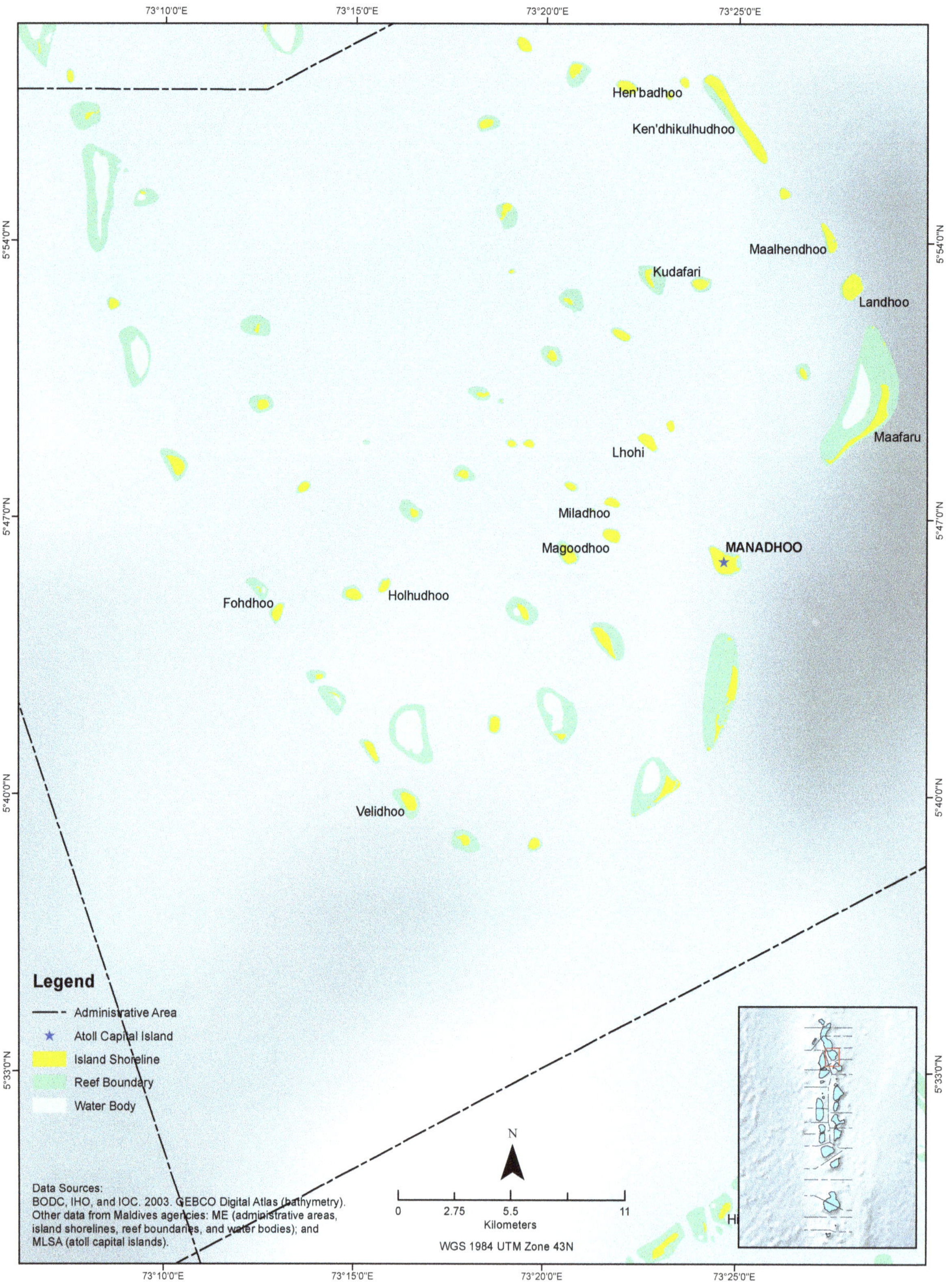
73°10'0"E
73°15'0"E
73°20'0"E
73°25'0"E
5°54'0"N
5°47'0"N
5°40'0"N
5°33'0"N
Hen'badhoo
Ken'dhikulhudhoo
Maalhendhoo
Kudafari
Landhoo
Maafaru
Lhohi
Miladhoo
Magoodhoo
MANADHOO
Fohdhoo
Holhudhoo
Velidhoo
Legend
Administrative Area
Atoll Capital Island
Island Shoreline
Reef Boundary
Water Body
N
0
2.75
5.5
11
Kilometers
WGS 1984 UTM Zone 43N

Data Sources:
BODC, IHO, and IOC. 2003. GEBCO Digital Atlas (bathymetry).
Other data from Maldives agencies: ME (administrative areas, island shorelines, reef boundaries, and water bodies); and MLSA (atoll capital islands).

Map I.18: North Malé Atoll, Basemap

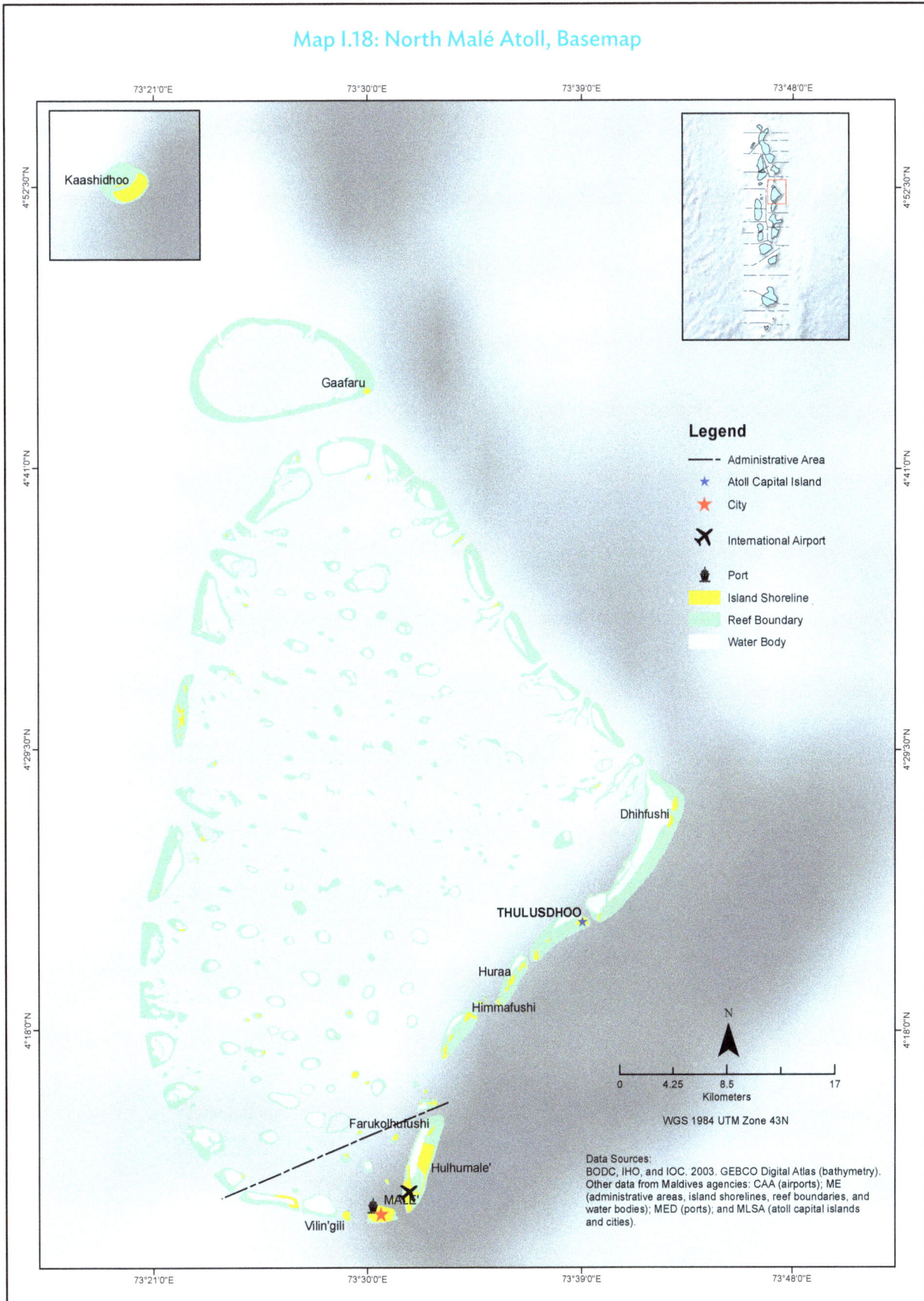

73°21'0"E   73°30'0"E   73°39'0"E   73°48'0"E

4°52'30"N

Kaashidhoo

4°41'0"N

Gaafaru

Legend

--- Administrative Area
★ Atoll Capital Island
★ City
✈ International Airport
⚓ Port
▮ Island Shoreline
▮ Reef Boundary
▮ Water Body

4°29'30"N

Dhihfushi

THULUSDHOO

Huraa

Himmafushi

N

0   4.25   8.5   17
Kilometers

WGS 1984 UTM Zone 43N

4°18'0"N

Farukolhufushi

Hulhumale'

MALE'

Vilin'gili

Data Sources:
BODC, IHO, and IOC. 2003. GEBCO Digital Atlas (bathymetry).
Other data from Maldives agencies: CAA (airports); ME
(administrative areas, island shorelines, reef boundaries, and
water bodies); MED (ports); and MLSA (atoll capital islands
and cities).

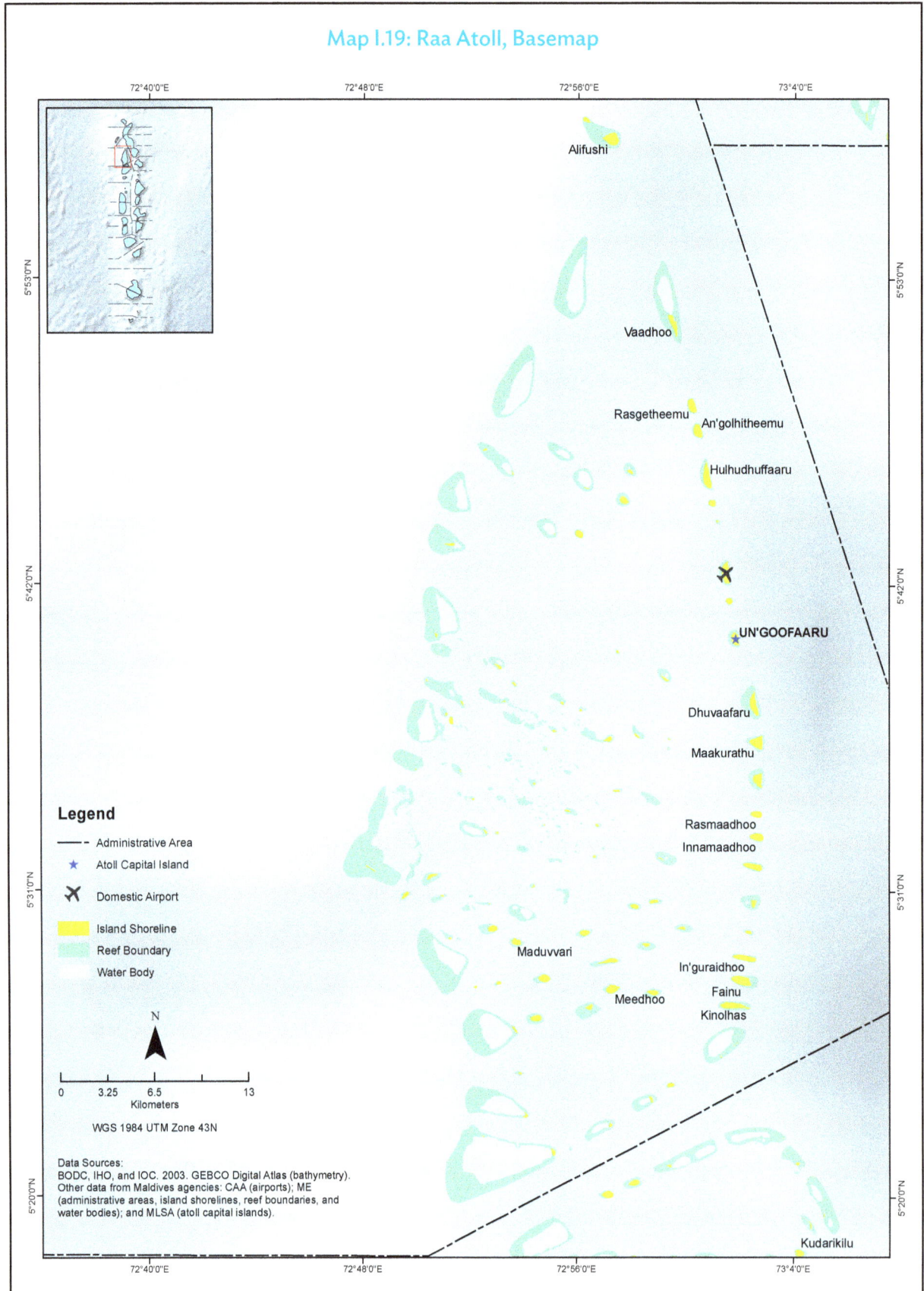

Map I.19: Raa Atoll, Basemap

72°40'0"E   72°48'0"E   72°56'0"E   73°4'0"E

Alifushi

Vaadhoo

Rasgetheemu   An'golhitheemu

Hulhudhuffaaru

UN'GOOFAARU

Dhuvaafaru

Maakurathu

Rasmaadhoo
Innamaadhoo

Legend

— · — Administrative Area

★ Atoll Capital Island

✈ Domestic Airport

▮ Island Shoreline

▮ Reef Boundary

     Water Body

N

Maduvvari

In'guraidhoo

Fainu

Meedhoo

Kinolhas

0   3.25   6.5   13
Kilometers

WGS 1984 UTM Zone 43N

Data Sources:
BODC, IHO, and IOC. 2003. GEBCO Digital Atlas (bathymetry).
Other data from Maldives agencies: CAA (airports); ME
(administrative areas, island shorelines, reef boundaries, and
water bodies); and MLSA (atoll capital islands).

Kudarikilu

72°40'0"E   72°48'0"E   72°56'0"E   73°4'0"E

5°53'0"N   5°42'0"N   5°31'0"N   5°20'0"N

Map I.20: Shaviyani Atoll, Basemap

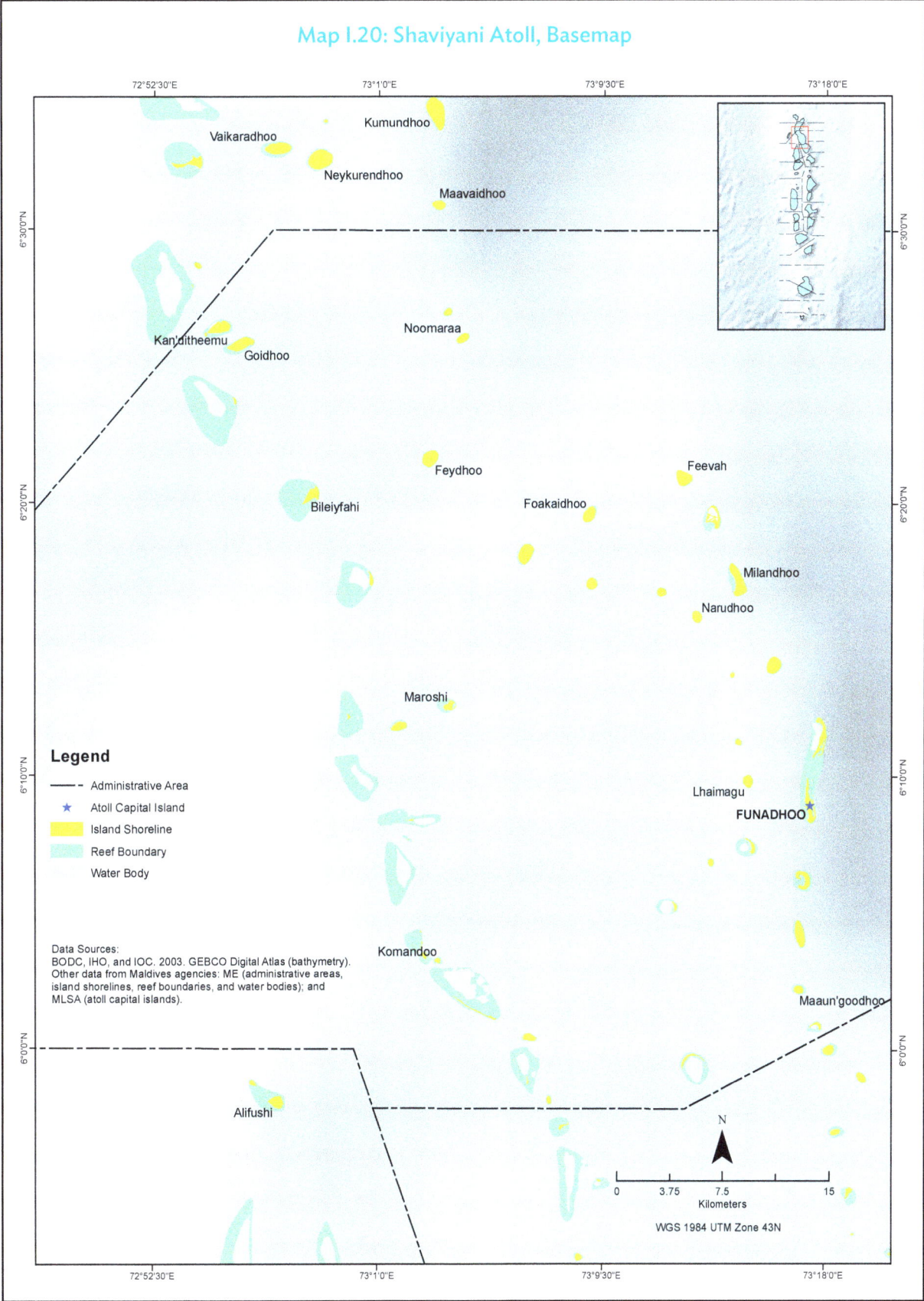

Kumundhoo

Vaikaradhoo

Neykurendhoo

Maavaidhoo

Noomaraa

Kan'ditheemu

Goidhoo

Feydhoo

Feevah

Bileiyfahi

Foakaidhoo

Milandhoo

Narudhoo

Maroshi

Lhaimagu

FUNADHOO

Legend

— Administrative Area

★ Atoll Capital Island

Island Shoreline

Reef Boundary

Water Body

Data Sources:
BODC, IHO, and IOC. 2003. GEBCO Digital Atlas (bathymetry).
Other data from Maldives agencies: ME (administrative areas,
island shorelines, reef boundaries, and water bodies); and
MLSA (atoll capital islands).

Komandoo

Maaun'goodhoo

Alifushi

N

0   3.75   7.5   15
Kilometers

WGS 1984 UTM Zone 43N

Map I.21: South Malé Atoll, Basemap

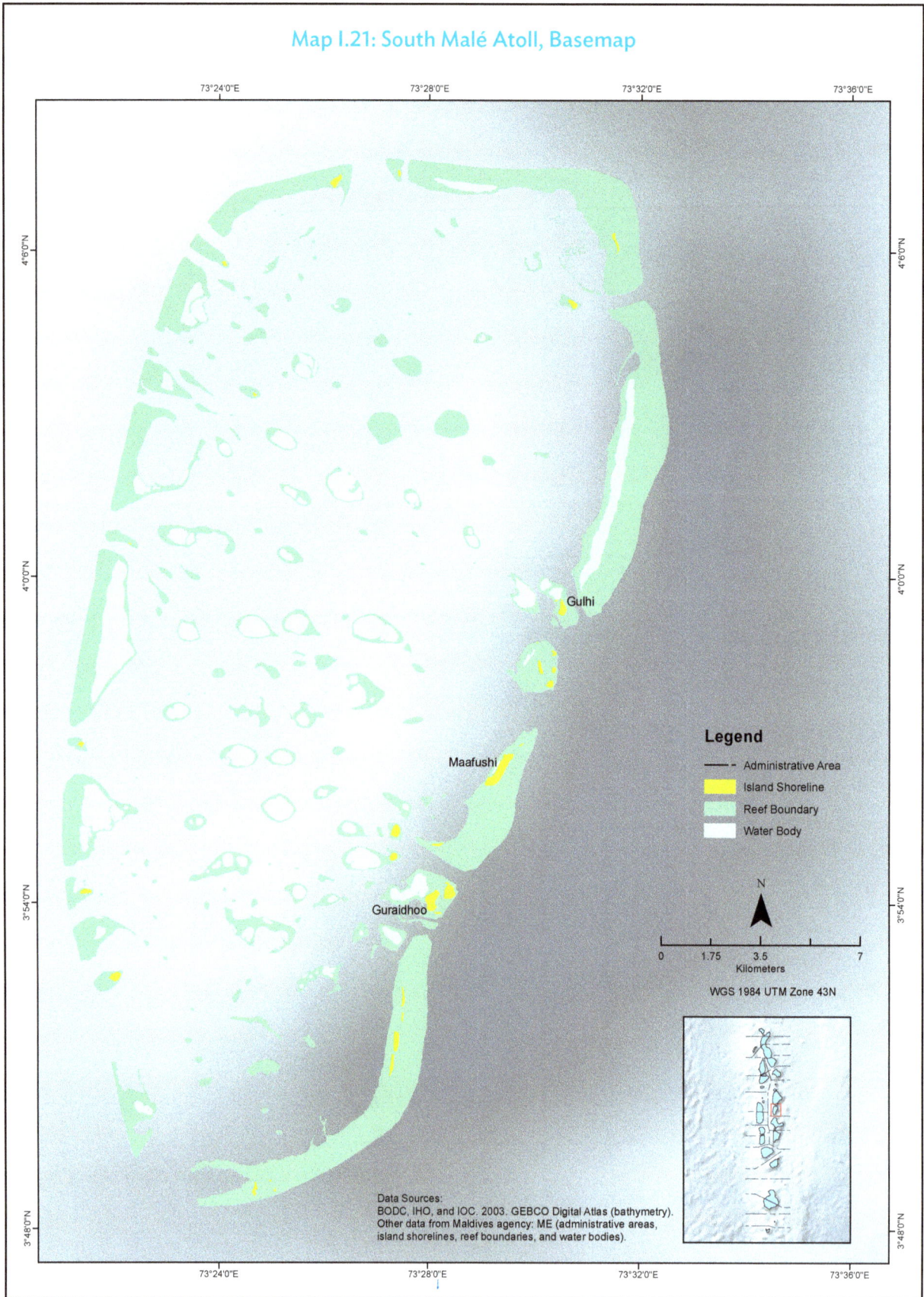

Gulhi

Maafushi

Guraidhoo

Legend

- - Administrative Area
- Island Shoreline
- Reef Boundary
- Water Body

N

0   1.75   3.5   7
Kilometers

WGS 1984 UTM Zone 43N

Data Sources:
BODC, IHO, and IOC. 2003. GEBCO Digital Atlas (bathymetry).
Other data from Maldives agency: ME (administrative areas,
island shorelines, reef boundaries, and water bodies).

Map I.22: Thaa Atoll, Basemap

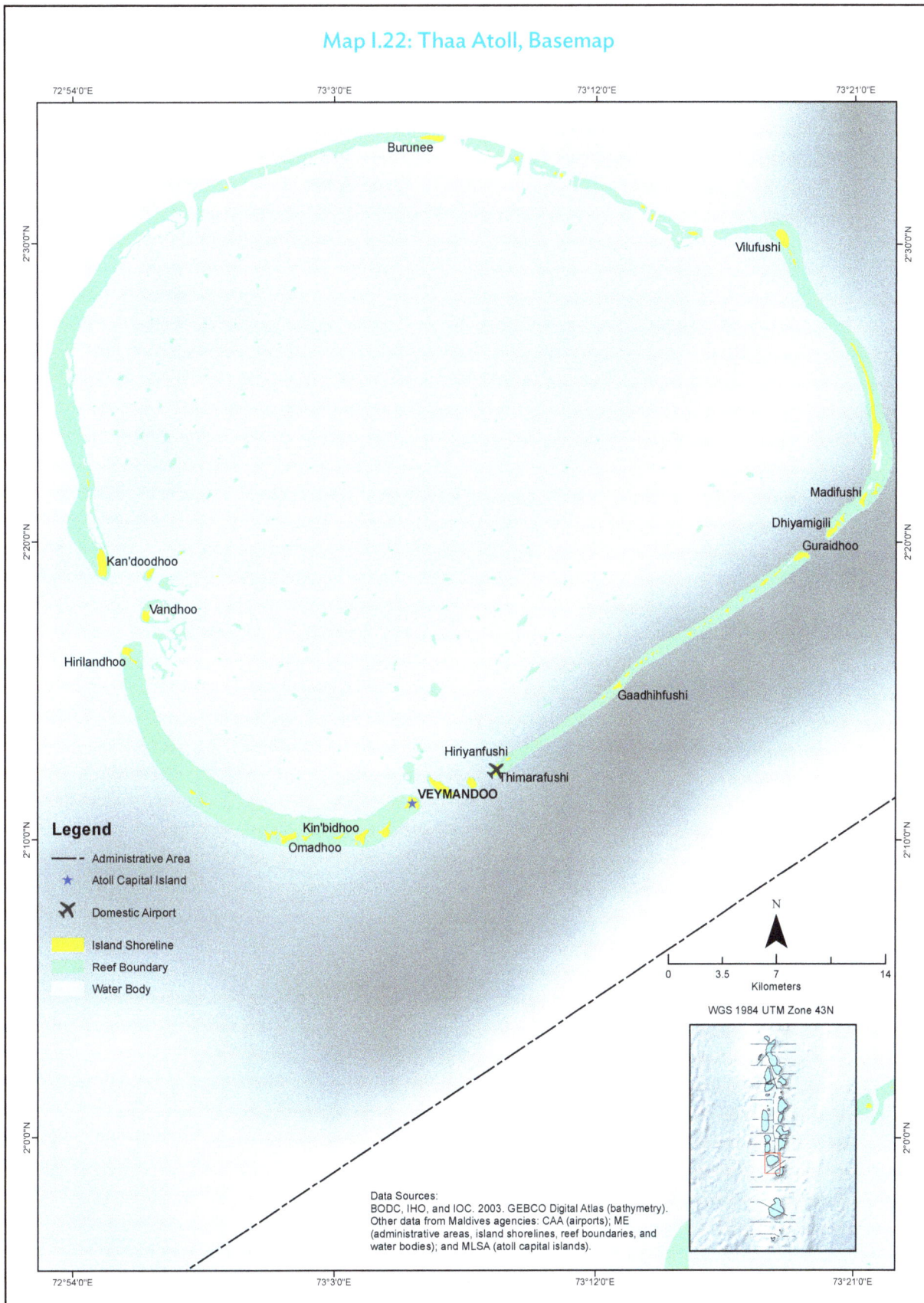

72°54'0"E   73°3'0"E   73°12'0"E   73°21'0"E

2°30'0"N

2°20'0"N

2°10'0"N

2°0'0"N

Burunee

Vilufushi

Madifushi

Dhiyamigili

Guraidhoo

Kan'doodhoo

Vandhoo

Hirilandhoo

Gaadhihfushi

Hiriyanfushi

Thimarafushi

VEYMANDOO

Kin'bidhoo

Omadhoo

Legend

— – Administrative Area

★ Atoll Capital Island

✈ Domestic Airport

▮ Island Shoreline

▮ Reef Boundary

▮ Water Body

N

0   3.5   7   14
Kilometers

WGS 1984 UTM Zone 43N

Data Sources:
BODC, IHO, and IOC. 2003. GEBCO Digital Atlas (bathymetry).
Other data from Maldives agencies: CAA (airports); ME
(administrative areas, island shorelines, reef boundaries, and
water bodies); and MLSA (atoll capital islands).

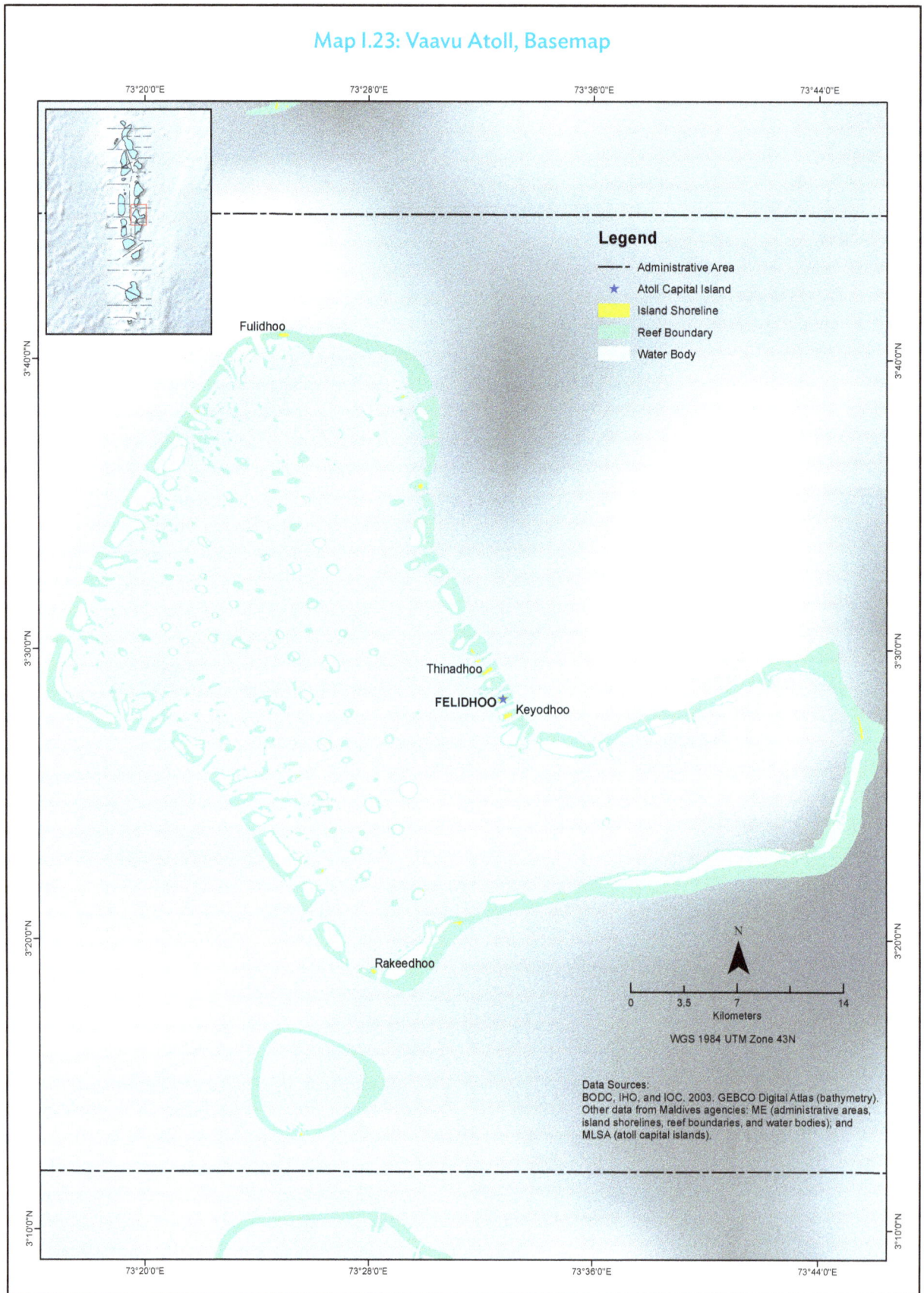

Map I.23: Vaavu Atoll, Basemap

73°20'0"E   73°28'0"E   73°36'0"E   73°44'0"E

Legend

------- Administrative Area

★ Atoll Capital Island

▣ Island Shoreline

▣ Reef Boundary

▢ Water Body

Fulidhoo

Thinadhoo

FELIDHOO ★ Keyodhoo

Rakeedhoo

3°40'0"N

3°30'0"N

3°20'0"N

3°10'0"N

N

0   3.5   7   14
Kilometers

WGS 1984 UTM Zone 43N

Data Sources:
BODC, IHO, and IOC. 2003. GEBCO Digital Atlas (bathymetry).
Other data from Maldives agencies: ME (administrative areas,
island shorelines, reef boundaries, and water bodies); and
MLSA (atoll capital islands).

Dynamic Geography

The natural geographic formation of Maldives presents challenges in urban expansion and development. Limited dry land combined with urban growth requires measures to maximize the available space for multiple activities. One mode of urban expansion is land reclamation. Land reclamation has become part of the development in the country and a solution to the limited available land, transforming the islands over the years to accommodate more human activities. Land reclamation is noticeable in the straightened coasts of Thinadhoo Island located in Gaafu Dhaalu Atoll (Map I.24). Other locations of land reclamation in Maldives are listed in Table I.1 and shown in Map I.25.

Table I.1: Islands with Reclamation

Atoll	No. of Islands with Reclamation	Islands with Reclamation
Haa Alifu	1	Dhidhoo
Haa Dhaalu	1	Kulhudhuffushi
Addu City	2	Hithadhoo
		Feydhoo
Alifu Alifu	2	Bodufolhudhoo
		Rasdhoo
Alifu Dhaalu	2	Dhihdhoo
		Maamigili
Faafu	2	Magoodhoo
		Nilandhoo
Laamu	2	Gaadhoo
		Kunahandhoo
Lhaviyani	2	Hinnavaru
		Naifaru
Noonu	2	Maafaru
		Vavathi
South Malé	2	Gulhi
		Maafushi

continued on next page

Table I.1 *continued*

Atoll	No. of Islands with Reclamation	Islands with Reclamation
Dhaalu	3	Kudahuvadhoo
		Maaen'boodhoo
		Meedhoo
Gaafu Alifu	3	Kolamaafushi
		Falhuverrahaa
		Dhaandhoo
Baa	4	Dharavandhoo
		Eydhafushi
		Thulhaadhoo
		Fares
Gaafu Dhaalu	4	Hoan'dedhdhoo
		Thinadhoo
		Faresmaathodaa
		Rodhavarrehaa
Meemu	4	Dhiggaru
		Maduhvari
		Muli
		Naalaafushi
Shaviyani	4	Milandhoo
		Funadhoo
		Komandoo
		Maroshi
Thaa	6	Vilufushi
		Madifushi
		Dhiyamigili
		Guraidhoo
		Hirilandhoo
		Thimarafushi
North Malé	8	Gaafaru
		Thulusdhoo
		Himmafushi
		Hulhumalé
		Malé
		Vilin'gili
		Hulhulé
		Tilafushi

Source: Government of Maldives, Ministry of National Planning and Infrastructure, 2017.

Land reclamation. To accommodate more people and urban activities, the coasts of major islands have been reclaimed. The Government of Maldives prioritized the Hulhumalé Land Reclamation and Development Project to address the need for more space around Malé City. This project will expand the island by 12.8 square kilometers in a sustainable manner (ADB 2017) (photo by Shahee Ilyas).

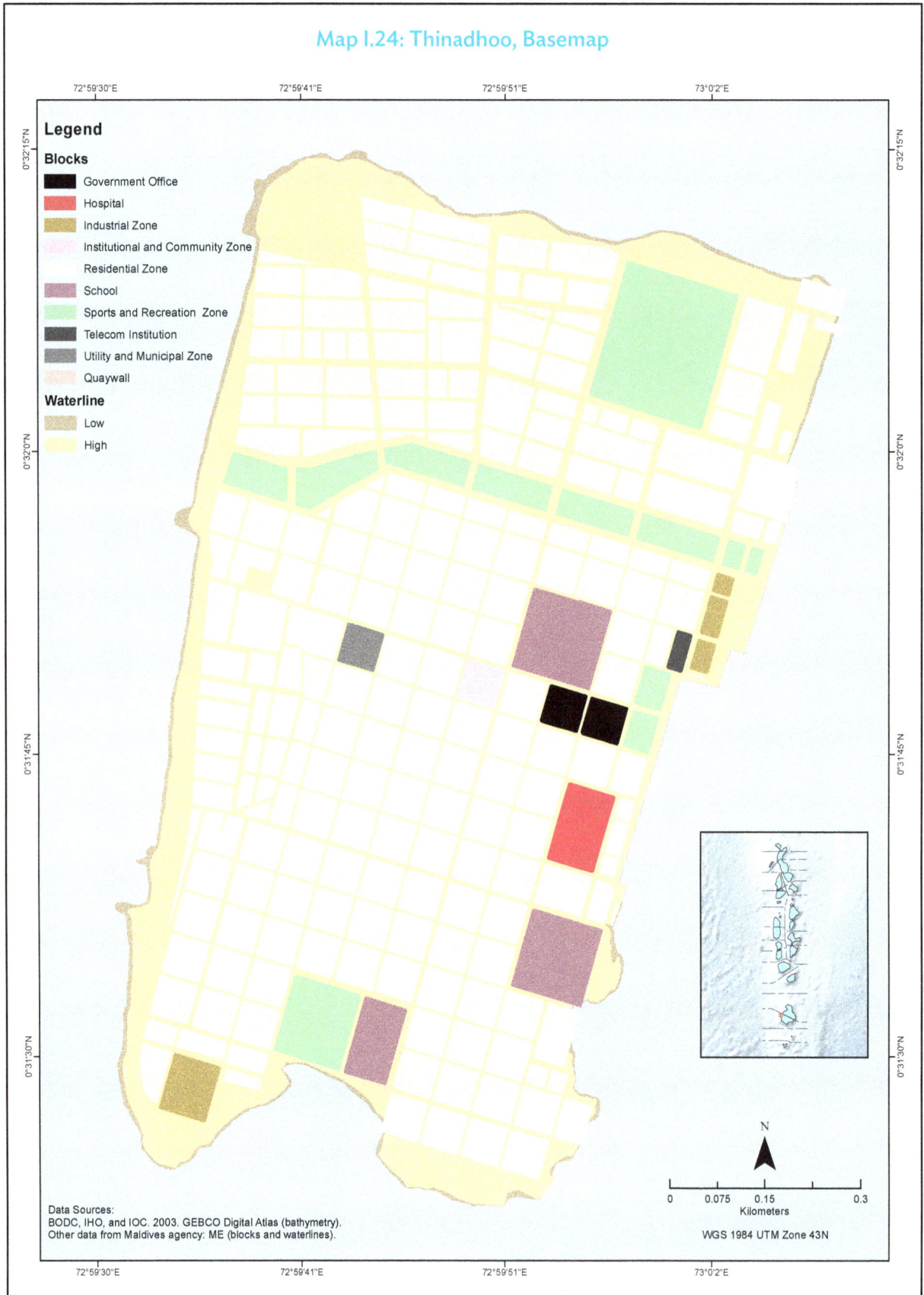

Map I.24: Thinadhoo, Basemap

72°59'30"E   72°59'41"E   72°59'51"E   73°0'2"E

0°32'15"N

Legend

Blocks
- ■ Government Office
- ■ Hospital
- ■ Industrial Zone
- ■ Institutional and Community Zone
- Residential Zone
- ■ School
- ■ Sports and Recreation Zone
- ■ Telecom Institution
- ■ Utility and Municipal Zone
- Quaywall

Waterline
- Low
- High

0°32'15"N

0°32'0"N

0°31'45"N

0°31'30"N

N

0   0.075   0.15   0.3
Kilometers

Data Sources:
BODC, IHO, and IOC. 2003. GEBCO Digital Atlas (bathymetry).
Other data from Maldives agency: ME (blocks and waterlines).

WGS 1984 UTM Zone 43N

72°59'30"E   72°59'41"E   72°59'51"E   73°0'2"E

Map I.25: Maldives, Land Reclamation

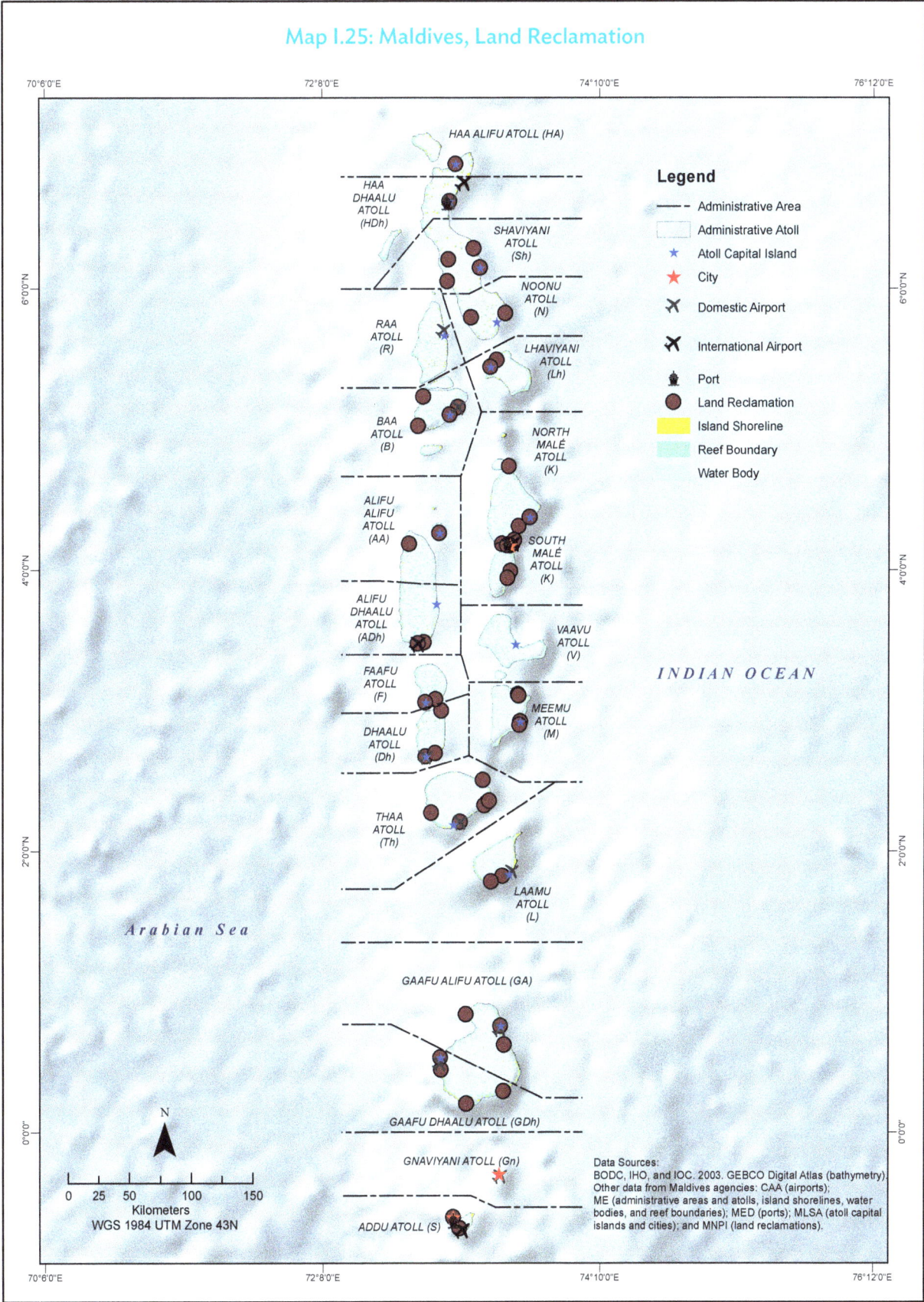

Legend

- — — Administrative Area
- ☐ Administrative Atoll
- ★ Atoll Capital Island
- ★ City
- ✈ Domestic Airport
- ✈ International Airport
- ⛴ Port
- ● Land Reclamation
- ▬ Island Shoreline
- ▬ Reef Boundary
- Water Body

INDIAN OCEAN

Arabian Sea

HAA ALIFU ATOLL (HA)
HAA DHAALU ATOLL (HDh)
SHAVIYANI ATOLL (Sh)
NOONU ATOLL (N)
RAA ATOLL (R)
LHAVIYANI ATOLL (Lh)
BAA ATOLL (B)
NORTH MALÉ ATOLL (K)
ALIFU ALIFU ATOLL (AA)
SOUTH MALÉ ATOLL (K)
ALIFU DHAALU ATOLL (ADh)
VAAVU ATOLL (V)
FAAFU ATOLL (F)
MEEMU ATOLL (M)
DHAALU ATOLL (Dh)
THAA ATOLL (Th)
LAAMU ATOLL (L)
GAAFU ALIFU ATOLL (GA)
GAAFU DHAALU ATOLL (GDh)
GNAVIYANI ATOLL (Gn)
ADDU ATOLL (S)

N

0 25 50   100   150
Kilometers
WGS 1984 UTM Zone 43N

Data Sources:
BODC, IHO, and IOC. 2003. GEBCO Digital Atlas (bathymetry).
Other data from Maldives agencies: CAA (airports);
ME (administrative areas and atolls, island shorelines, water
bodies, and reef boundaries); MED (ports); MLSA (atoll capital
islands and cities); and MNPI (land reclamations).

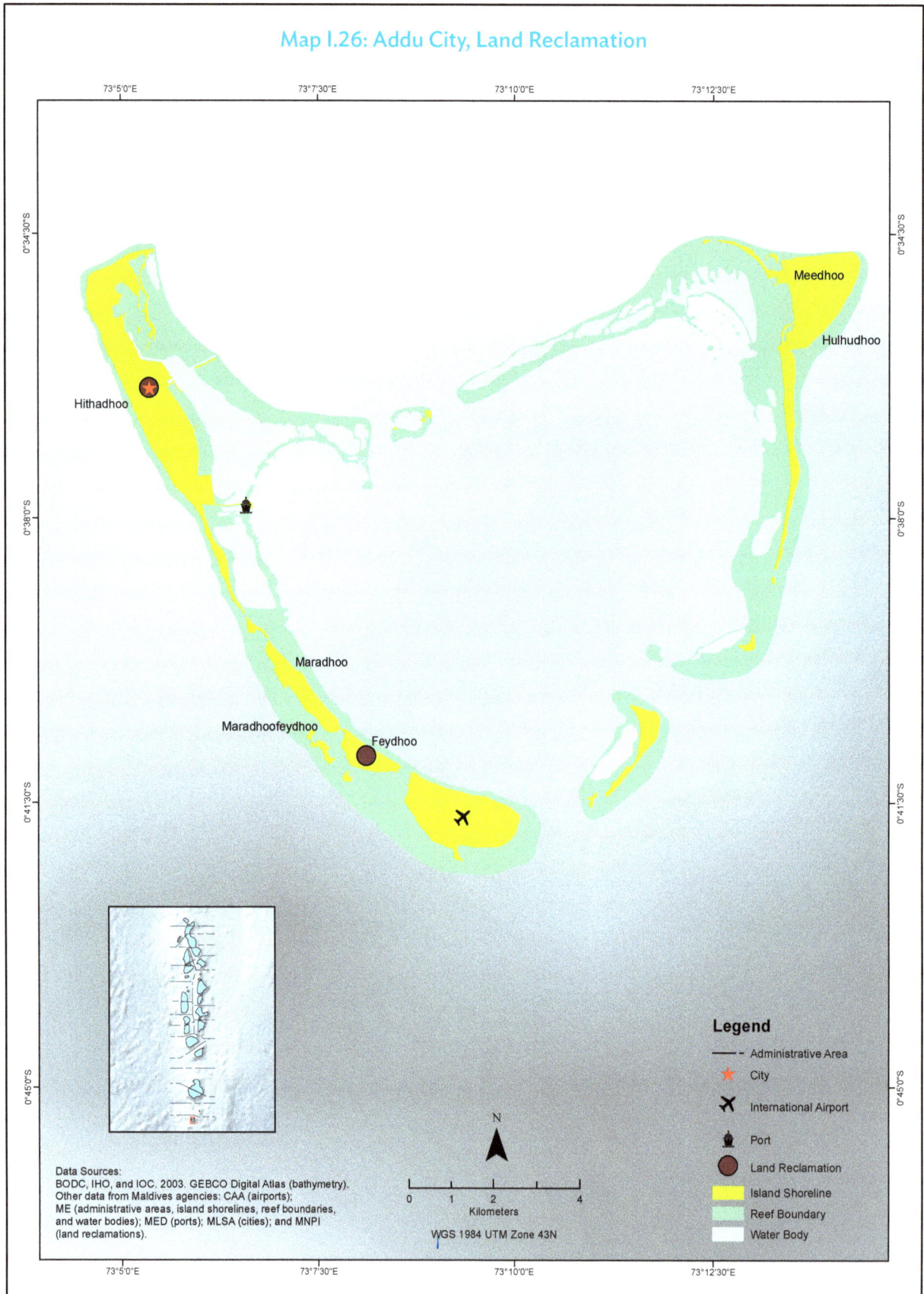

Map I.26: Addu City, Land Reclamation

Meedhoo

Hulhudhoo

Hithadhoo

Maradhoo

Maradhoofeydhoo

Feydhoo

Data Sources:
BODC, IHO, and IOC. 2003. GEBCO Digital Atlas (bathymetry).
Other data from Maldives agencies: CAA (airports);
ME (administrative areas, island shorelines, reef boundaries,
and water bodies); MED (ports); MLSA (cities); and MNPI
(land reclamations).

N

0   1   2   4
Kilometers

WGS 1984 UTM Zone 43N

Legend

— - Administrative Area
★ City
✗ International Airport
⚓ Port
● Land Reclamation
▮ Island Shoreline
▮ Reef Boundary
▯ Water Body

Map I.27: Alifu Alifu Atoll, Land Reclamation

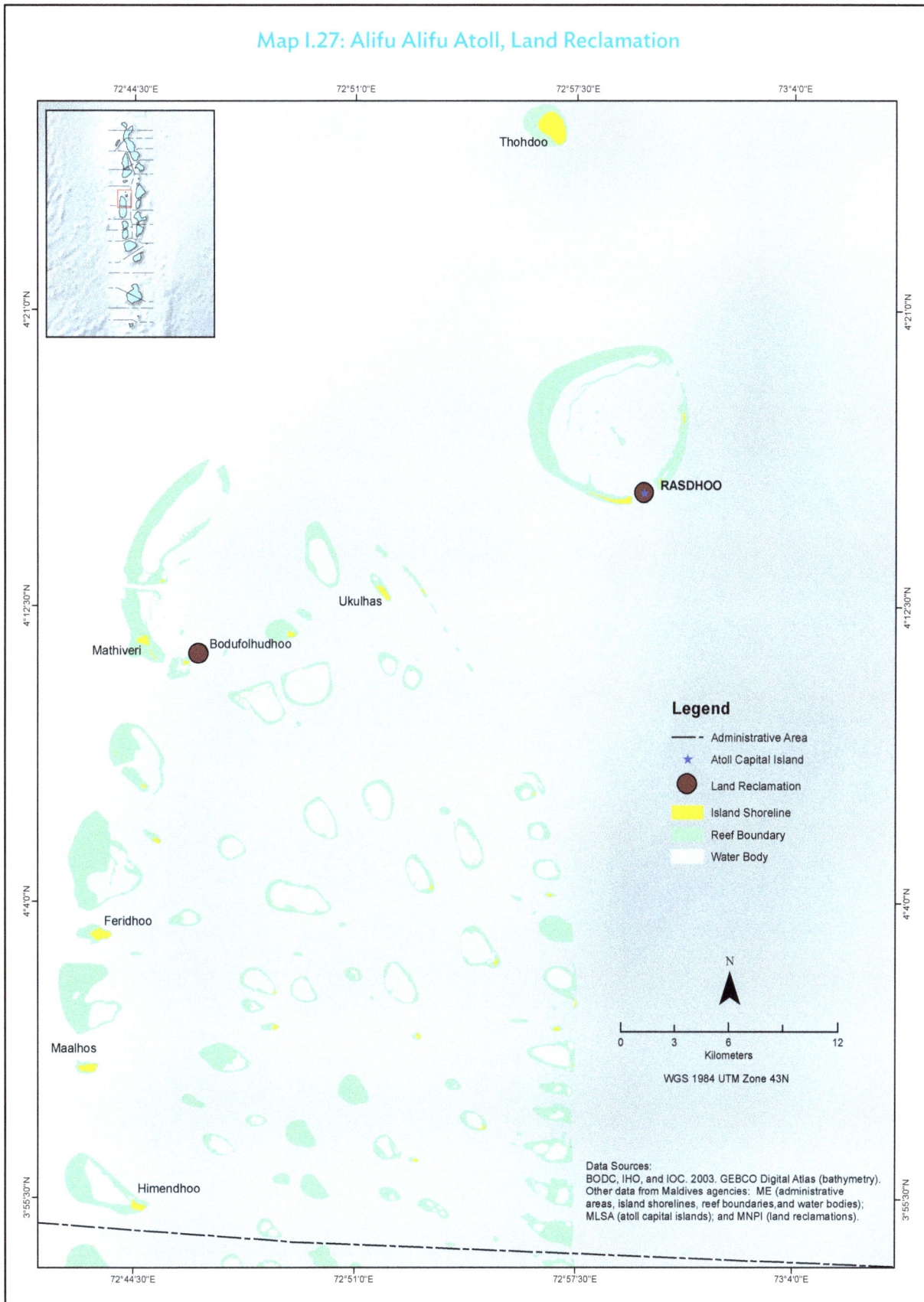

Thohdoo

RASDHOO

Ukulhas

Mathiveri   Bodufolhudhoo

Legend

— Administrative Area
★ Atoll Capital Island
● Land Reclamation
▮ Island Shoreline
▮ Reef Boundary
▯ Water Body

Feridhoo

Maalhos

N

0   3   6   12
Kilometers
WGS 1984 UTM Zone 43N

Himendhoo

Data Sources:
BODC, IHO, and IOC. 2003. GEBCO Digital Atlas (bathymetry).
Other data from Maldives agencies: ME (administrative
areas, island shorelines, reef boundaries,and water bodies);
MLSA (atoll capital islands); and MNPI (land reclamations).

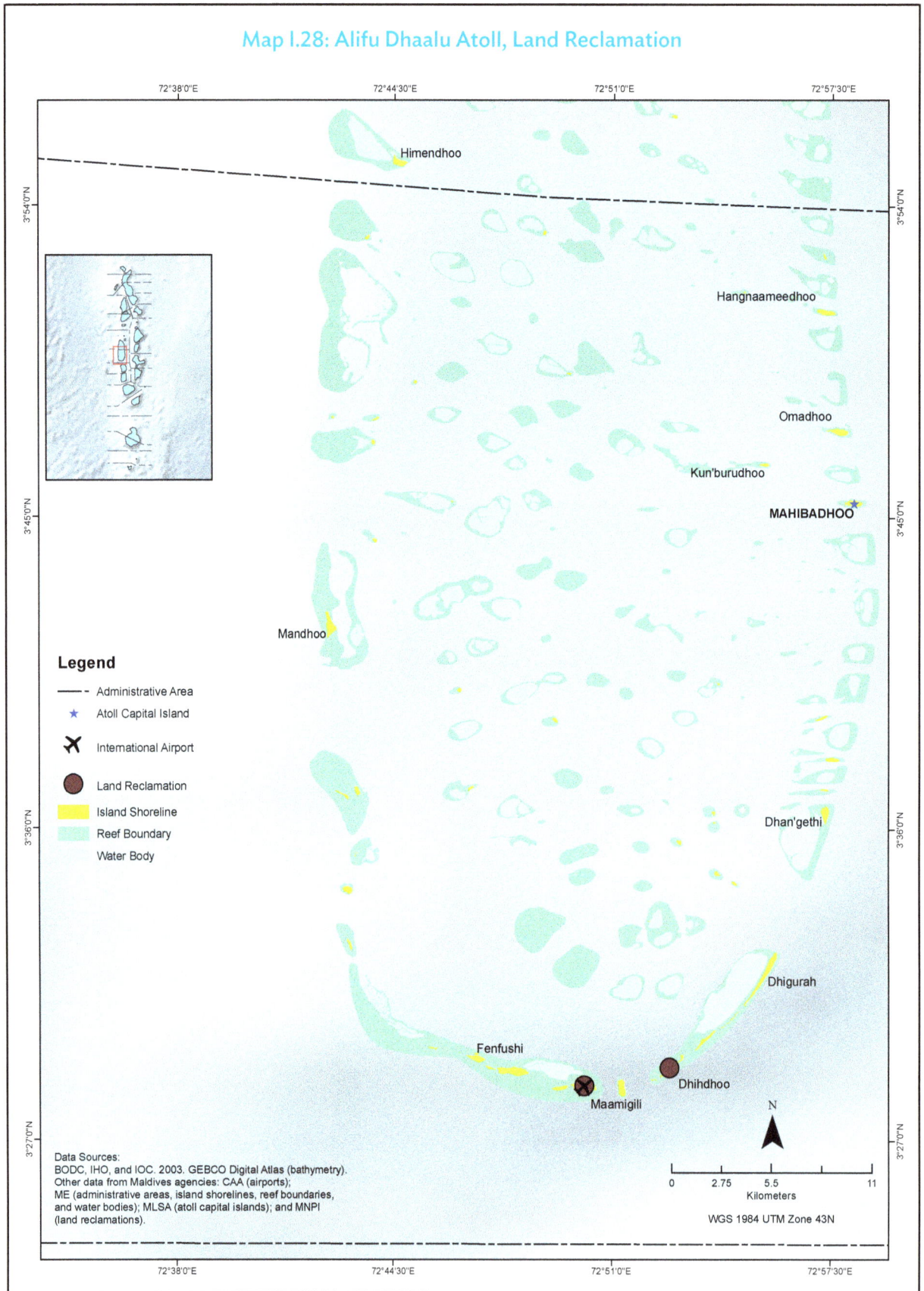

Map I.28: Alifu Dhaalu Atoll, Land Reclamation

72°38'0"E   72°44'30"E   72°51'0"E   72°57'30"E

3°54'0"N

Himendhoo

Hangnaameedhoo

Omadhoo

3°45'0"N

Kun'burudhoo

MAHIBADHOO ★

Mandhoo

Legend

— · — Administrative Area

★ Atoll Capital Island

✈ International Airport

⬤ Land Reclamation

▮ Island Shoreline

▮ Reef Boundary

Water Body

Dhan'gethi

3°36'0"N

Dhigurah

Fenfushi

Dhihdhoo

Maamigili

N

3°27'0"N

Data Sources:
BODC, IHO, and IOC. 2003. GEBCO Digital Atlas (bathymetry).
Other data from Maldives agencies: CAA (airports);
ME (administrative areas, island shorelines, reef boundaries,
and water bodies); MLSA (atoll capital islands); and MNPI
(land reclamations).

0   2.75   5.5   11
Kilometers

WGS 1984 UTM Zone 43N

72°38'0"E   72°44'30"E   72°51'0"E   72°57'30"E

Map I.29: Baa Atoll, Land Reclamation

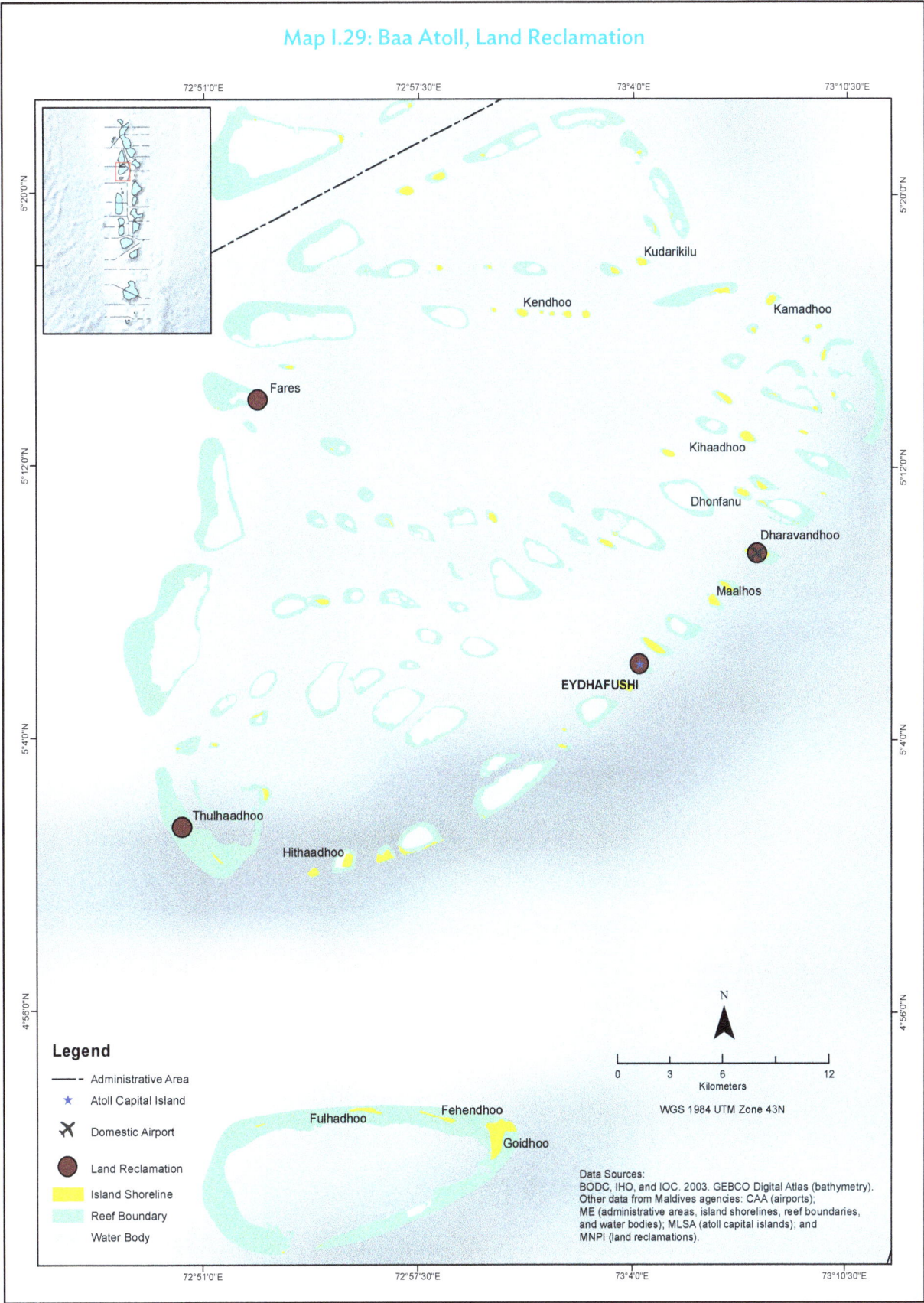

Kudarikilu

Kendhoo

Kamadhoo

Fares

Kihaadhoo

Dhonfanu

Dharavandhoo

Maalhos

EYDHAFUSHI

Thulhaadhoo

Hithaadhoo

N

0   3   6   12
Kilometers

WGS 1984 UTM Zone 43N

Fehendhoo

Fulhadhoo

Goidhoo

Legend

- – – Administrative Area
- ★ Atoll Capital Island
- ✈ Domestic Airport
- ● Land Reclamation
- ▮ Island Shoreline
- ▮ Reef Boundary
- Water Body

Data Sources:
BODC, IHO, and IOC. 2003. GEBCO Digital Atlas (bathymetry).
Other data from Maldives agencies: CAA (airports);
ME (administrative areas, island shorelines, reef boundaries,
and water bodies); MLSA (atoll capital islands); and
MNPI (land reclamations).

Map I.30: Dhaalu Atoll, Land Reclamation

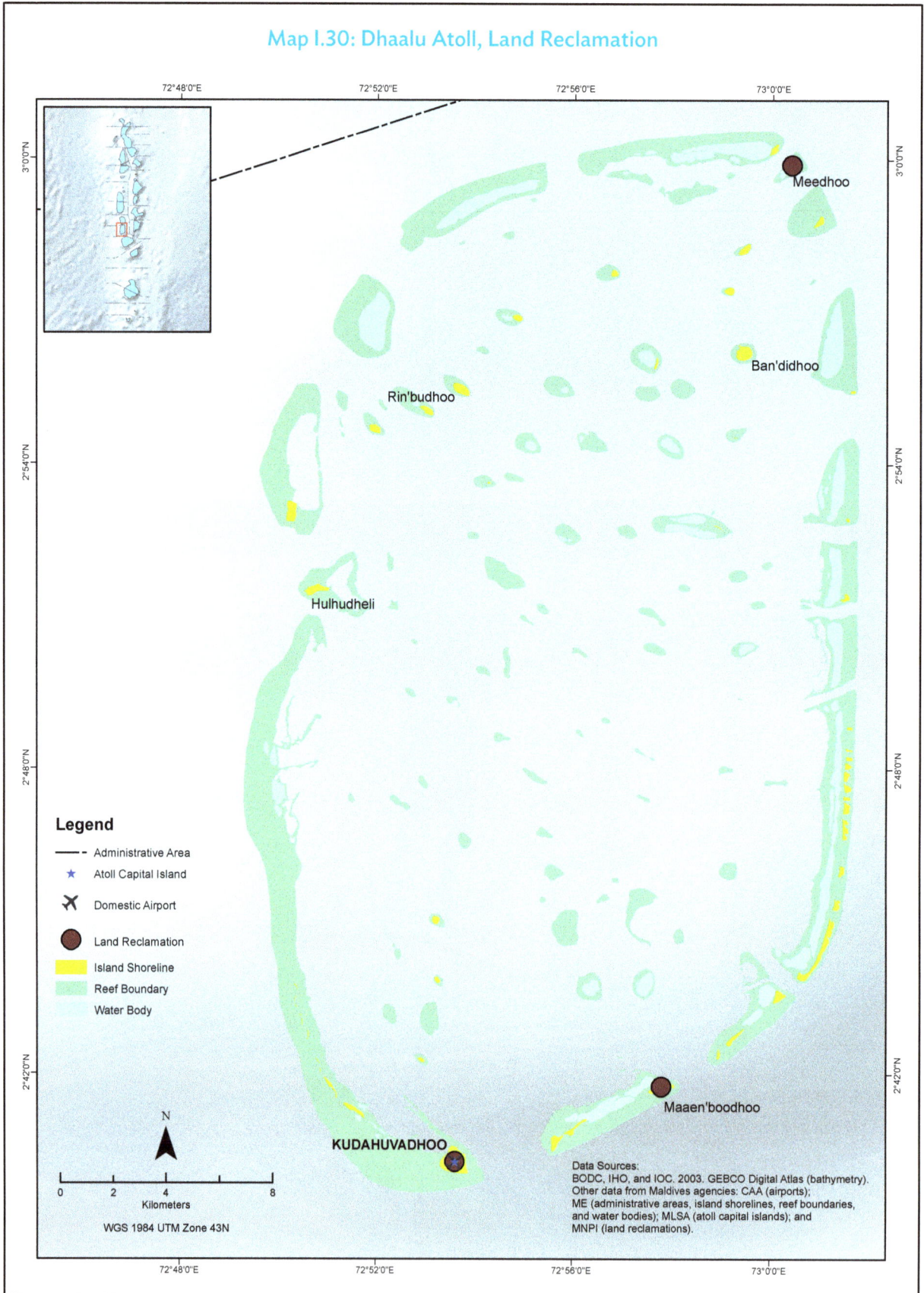

Meedhoo

Ban'didhoo

Rin'budhoo

Hulhudheli

Legend

— - — Administrative Area

★ Atoll Capital Island

✈ Domestic Airport

⬤ Land Reclamation

▮ Island Shoreline

▮ Reef Boundary

  Water Body

N

0   2   4   8
Kilometers

WGS 1984 UTM Zone 43N

KUDAHUVADHOO

Maaen'boodhoo

Data Sources:
BODC, IHO, and IOC. 2003. GEBCO Digital Atlas (bathymetry).
Other data from Maldives agencies: CAA (airports);
ME (administrative areas, island shorelines, reef boundaries,
and water bodies); MLSA (atoll capital islands); and
MNPI (land reclamations).

Map I.31: Faafu Atoll, Land Reclamation

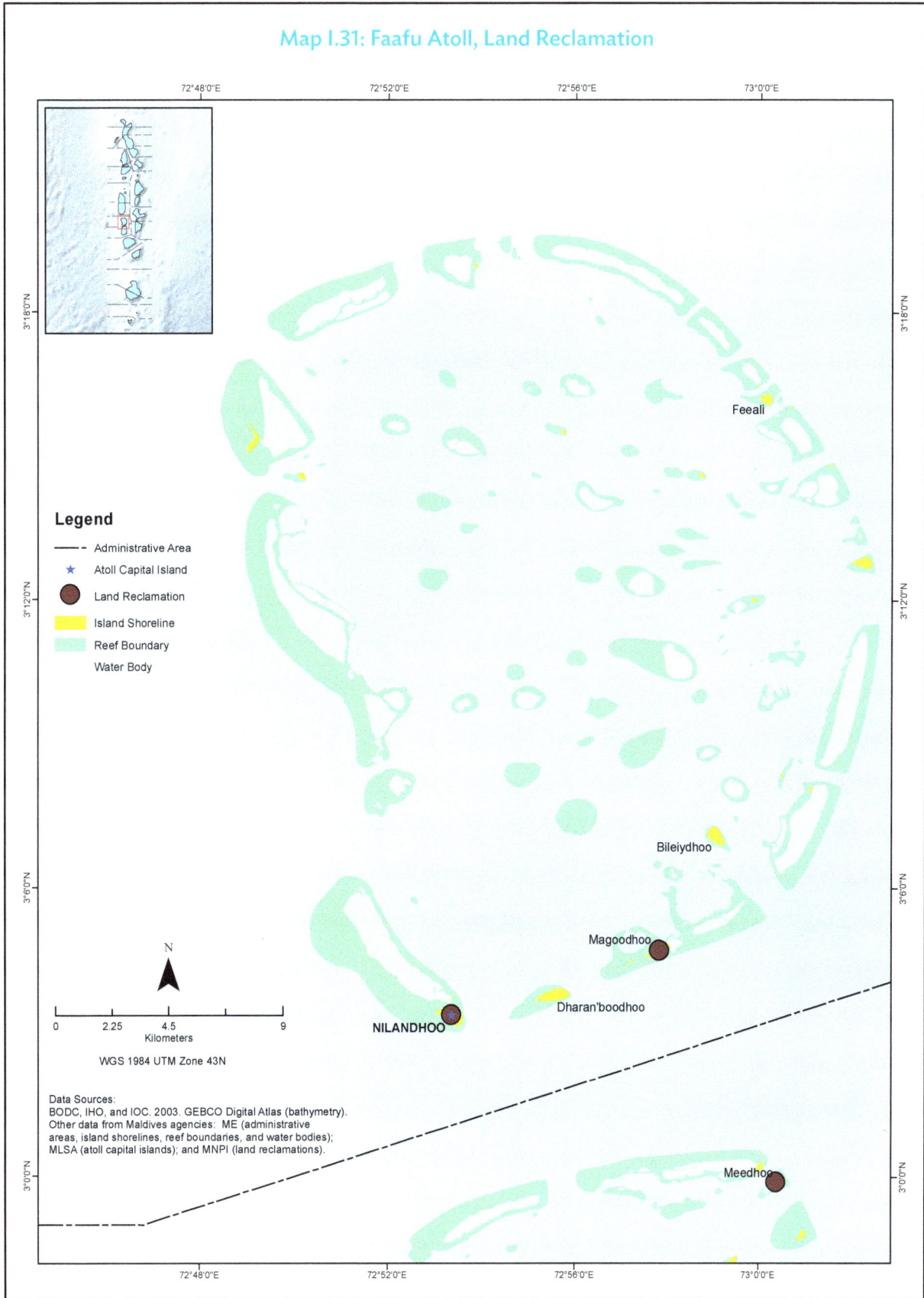

72°48'0"E   72°52'0"E   72°56'0"E   73°0'0"E

3°18'0"N

Feeali

Legend

--- Administrative Area

★ Atoll Capital Island

⬤ Land Reclamation

Island Shoreline

Reef Boundary

Water Body

3°12'0"N

Bileiydhoo

3°6'0"N

Magoodhoo

N

Dharan'boodhoo

0   2.25   4.5   9
Kilometers

NILANDHOO

WGS 1984 UTM Zone 43N

Data Sources:
BODC, IHO, and IOC. 2003. GEBCO Digital Atlas (bathymetry).
Other data from Maldives agencies: ME (administrative
areas, island shorelines, reef boundaries, and water bodies);
MLSA (atoll capital islands); and MNPI (land reclamations).

Meedhoo

3°0'0"N

72°48'0"E   72°52'0"E   72°56'0"E   73°0'0"E

Map I.32: Gaafu Alifu Atoll, Land Reclamation

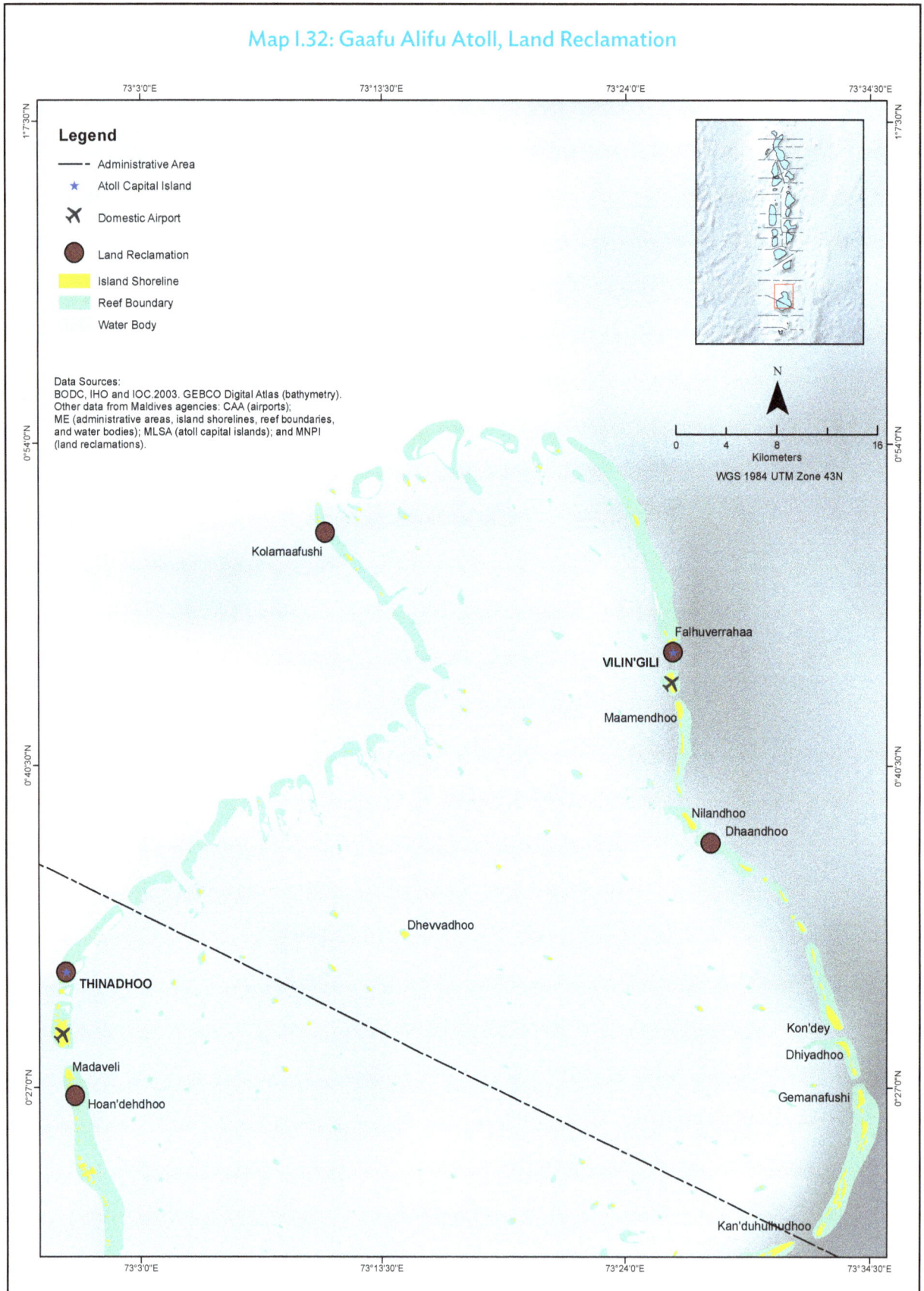

Legend

— — Administrative Area

★ Atoll Capital Island

✈ Domestic Airport

● Land Reclamation

▮ Island Shoreline

▮ Reef Boundary

▢ Water Body

Data Sources:
BODC, IHO and IOC.2003. GEBCO Digital Atlas (bathymetry).
Other data from Maldives agencies: CAA (airports);
ME (administrative areas, island shorelines, reef boundaries,
and water bodies); MLSA (atoll capital islands); and MNPI
(land reclamations).

N

0   4   8   16
Kilometers
WGS 1984 UTM Zone 43N

Kolamaafushi

Falhuverrahaa
VILIN'GILI
Maamendhoo

Nilandhoo
Dhaandhoo

Dhevvadhoo

THINADHOO

Kon'dey
Dhiyadhoo

Madaveli
Gemanafushi
Hoan'dehdhoo

Kan'duhulhudhoo

Map I.33: Gaafu Dhaalu Atoll, Land Reclamation

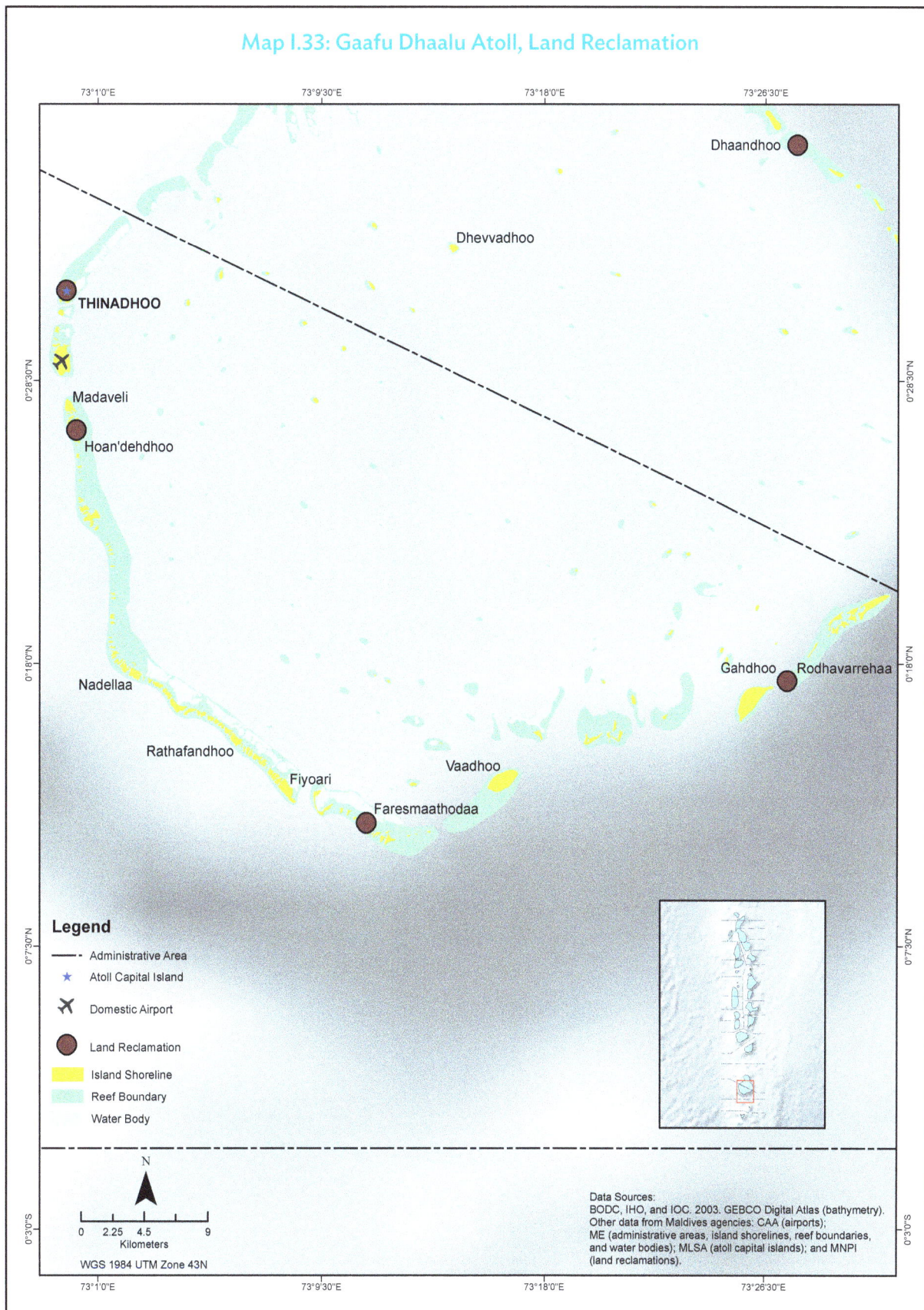

73°1'0"E   73°9'30"E   73°18'0"E   73°26'30"E

Dhaandhoo

Dhevvadhoo

THINADHOO

Madaveli

Hoan'dehdhoo

Nadellaa

Gahdhoo   Rodhavarrehaa

Rathafandhoo

Fiyoari

Vaadhoo

Faresmaathodaa

Legend

— · — Administrative Area

★ Atoll Capital Island

✈ Domestic Airport

● Land Reclamation

▮ Island Shoreline

▮ Reef Boundary

Water Body

N

0   2.25   4.5   9
Kilometers

WGS 1984 UTM Zone 43N

Data Sources:
BODC, IHO, and IOC. 2003. GEBCO Digital Atlas (bathymetry).
Other data from Maldives agencies: CAA (airports);
ME (administrative areas, island shorelines, reef boundaries,
and water bodies); MLSA (atoll capital islands); and MNPI
(land reclamations).

Map I.34: Haa Alifu Atoll, Land Reclamation

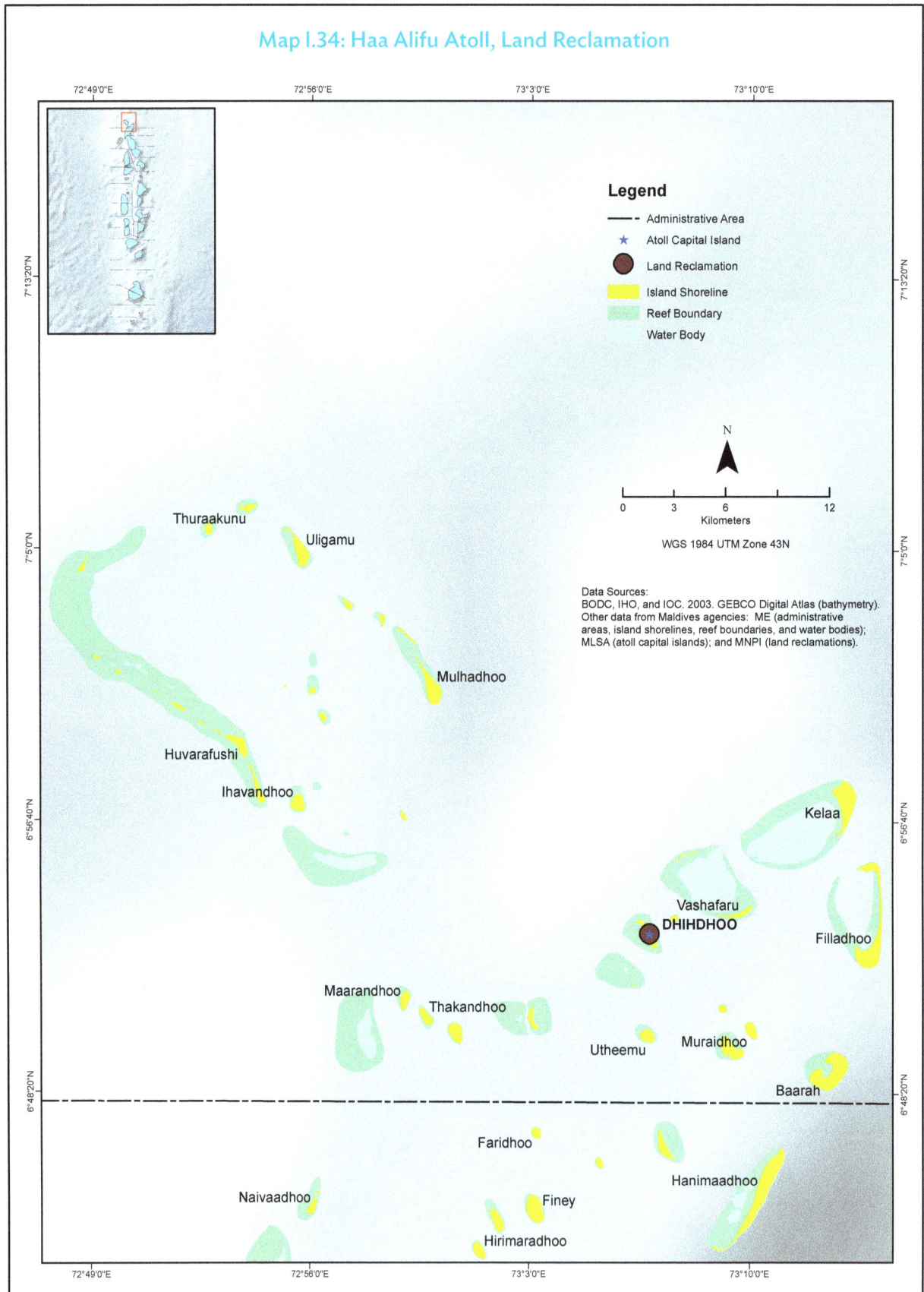

Legend

- — Administrative Area
- ★ Atoll Capital Island
- ● Land Reclamation
- ▨ Island Shoreline
- ▨ Reef Boundary
- Water Body

N

0   3   6   12
Kilometers

WGS 1984 UTM Zone 43N

Data Sources:
BODC, IHO, and IOC. 2003. GEBCO Digital Atlas (bathymetry).
Other data from Maldives agencies: ME (administrative
areas, island shorelines, reef boundaries, and water bodies);
MLSA (atoll capital islands); and MNPI (land reclamations).

Thuraakunu
Uligamu
Mulhadhoo
Huvarafushi
Ihavandhoo
Kelaa
Vashafaru
DHIHDHOO
Filladhoo
Maarandhoo
Thakandhoo
Muraidhoo
Utheemu
Baarah
Faridhoo
Hanimaadhoo
Naivaadhoo
Finey
Hirimaradhoo

Map I.35: Haa Dhaalu Atoll, Land Reclamation

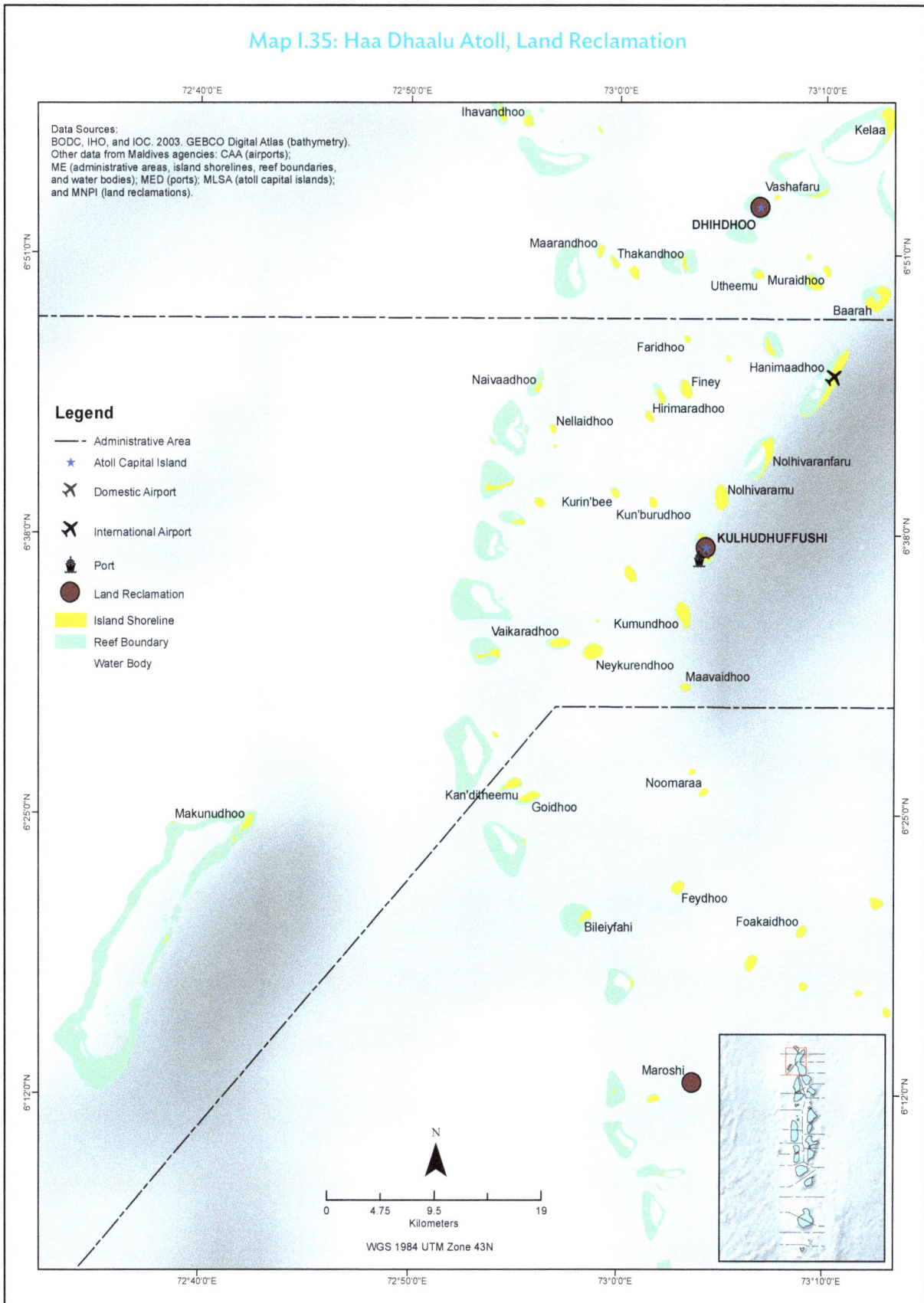

72°40'0"E   72°50'0"E   73°0'0"E   73°10'0"E

Data Sources:
BODC, IHO, and IOC. 2003. GEBCO Digital Atlas (bathymetry).
Other data from Maldives agencies: CAA (airports);
ME (administrative areas, island shorelines, reef boundaries,
and water bodies); MED (ports); MLSA (atoll capital islands);
and MNPI (land reclamations).

Ihavandhoo

Kelaa

Vashafaru

DHIHDHOO

Maarandhoo   Thakandhoo

Utheemu   Muraidhoo

Baarah

Faridhoo

Naivaadhoo   Hanimaadhoo

Finey

Hirimaradhoo

Nellaidhoo

Nolhivaranfaru

Nolhivaramu

Kurin'bee   Kun'burudhoo

KULHUDHUFFUSHI

6°51'0"N

6°38'0"N

6°25'0"N

6°12'0"N

Legend

– – – Administrative Area

★ Atoll Capital Island

✈ Domestic Airport

✈ International Airport

⚓ Port

● Land Reclamation

▮ Island Shoreline

▮ Reef Boundary

   Water Body

Kumundhoo

Vaikaradhoo

Neykurendhoo   Maavaidhoo

Noomaraa

Makunudhoo

Kan'ditheemu   Goidhoo

Feydhoo

Bileiyfahi   Foakaidhoo

Maroshi

N

0   4.75   9.5   19
Kilometers

WGS 1984 UTM Zone 43N

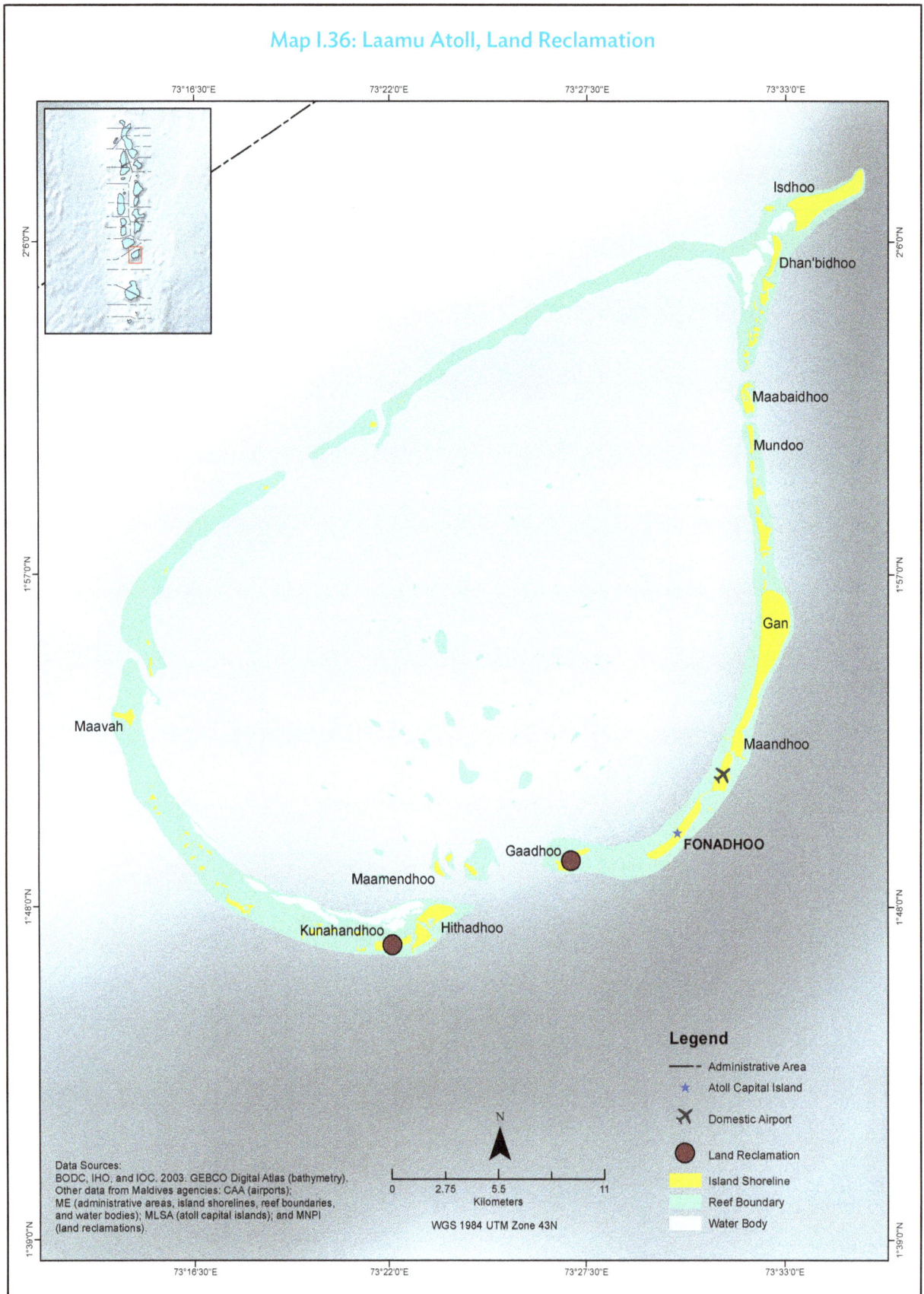

Map I.36: Laamu Atoll, Land Reclamation

73°16'30"E   73°22'0"E   73°27'30"E   73°33'0"E

2°6'0"N

Isdhoo

Dhan'bidhoo

Maabaidhoo

Mundoo

1°57'0"N

Gan

Maavah

Maandhoo

FONADHOO

Gaadhoo

Maamendhoo

1°48'0"N

Kunahandhoo   Hithadhoo

Legend

- – – Administrative Area
- ★ Atoll Capital Island
- ✈ Domestic Airport
- ● Land Reclamation
- ▮ Island Shoreline
- ▮ Reef Boundary
- ▮ Water Body

Data Sources:
BODC, IHO, and IOC. 2003. GEBCO Digital Atlas (bathymetry).
Other data from Maldives agencies: CAA (airports);
ME (administrative areas, island shorelines, reef boundaries,
and water bodies); MLSA (atoll capital islands); and MNPI
(land reclamations).

N

0   2.75   5.5   11
Kilometers

WGS 1984 UTM Zone 43N

Map I.37: Lhaviyani Atoll, Land Reclamation

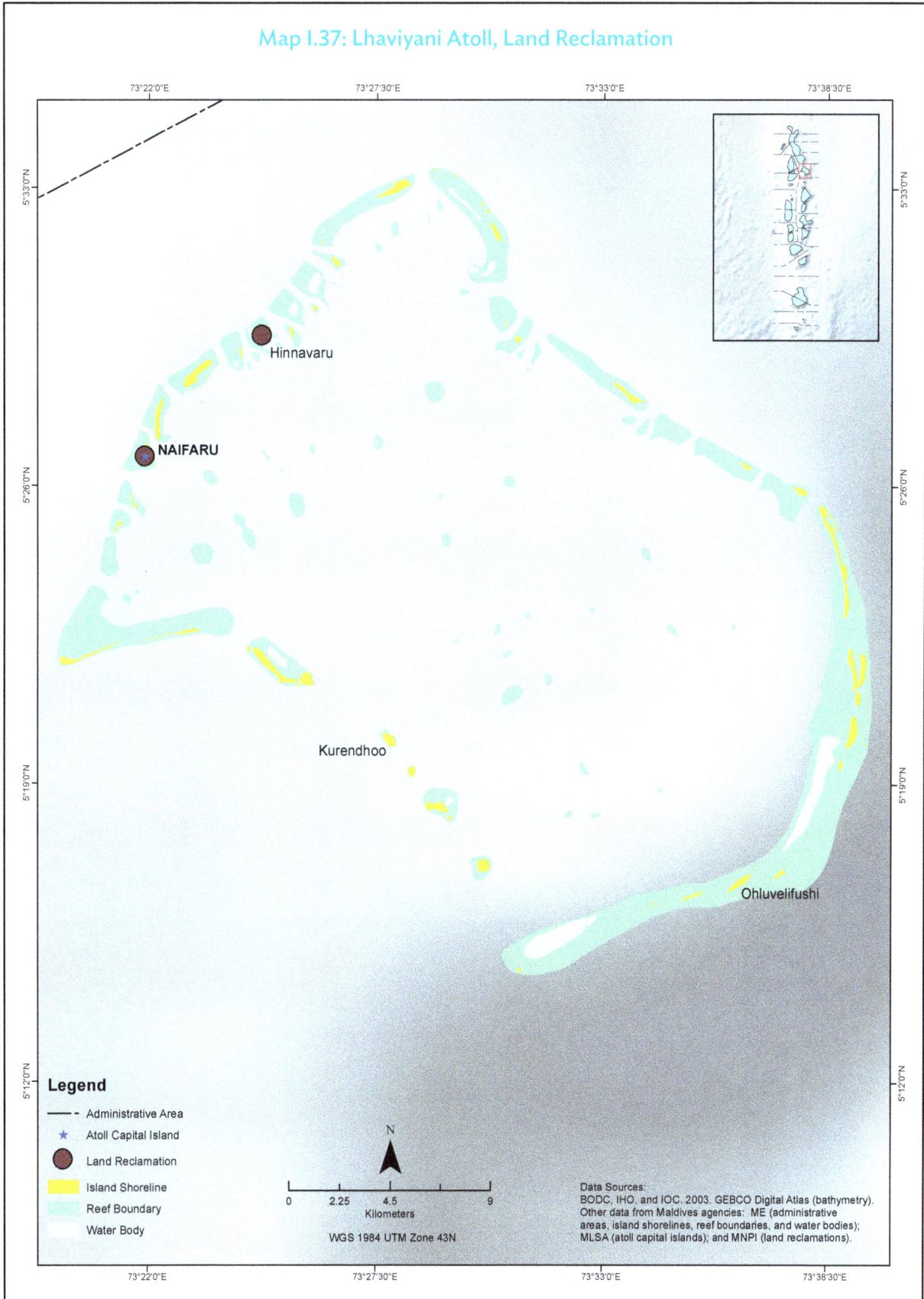

73°22'0"E   73°27'30"E   73°33'0"E   73°38'30"E

5°33'0"N

Hinnavaru

NAIFARU

5°26'0"N

Kurendhoo

5°19'0"N

Ohluvelifushi

5°12'0"N

Legend

— · — Administrative Area

★ Atoll Capital Island

● Land Reclamation

Island Shoreline

Reef Boundary

Water Body

N

0   2.25   4.5   9
Kilometers
WGS 1984 UTM Zone 43N

Data Sources:
BODC, IHO, and IOC. 2003. GEBCO Digital Atlas (bathymetry).
Other data from Maldives agencies: ME (administrative
areas, island shorelines, reef boundaries, and water bodies);
MLSA (atoll capital islands); and MNPI (land reclamations).

73°22'0"E   73°27'30"E   73°33'0"E   73°38'30"E

Map I.38: Meemu Atoll, Land Reclamation

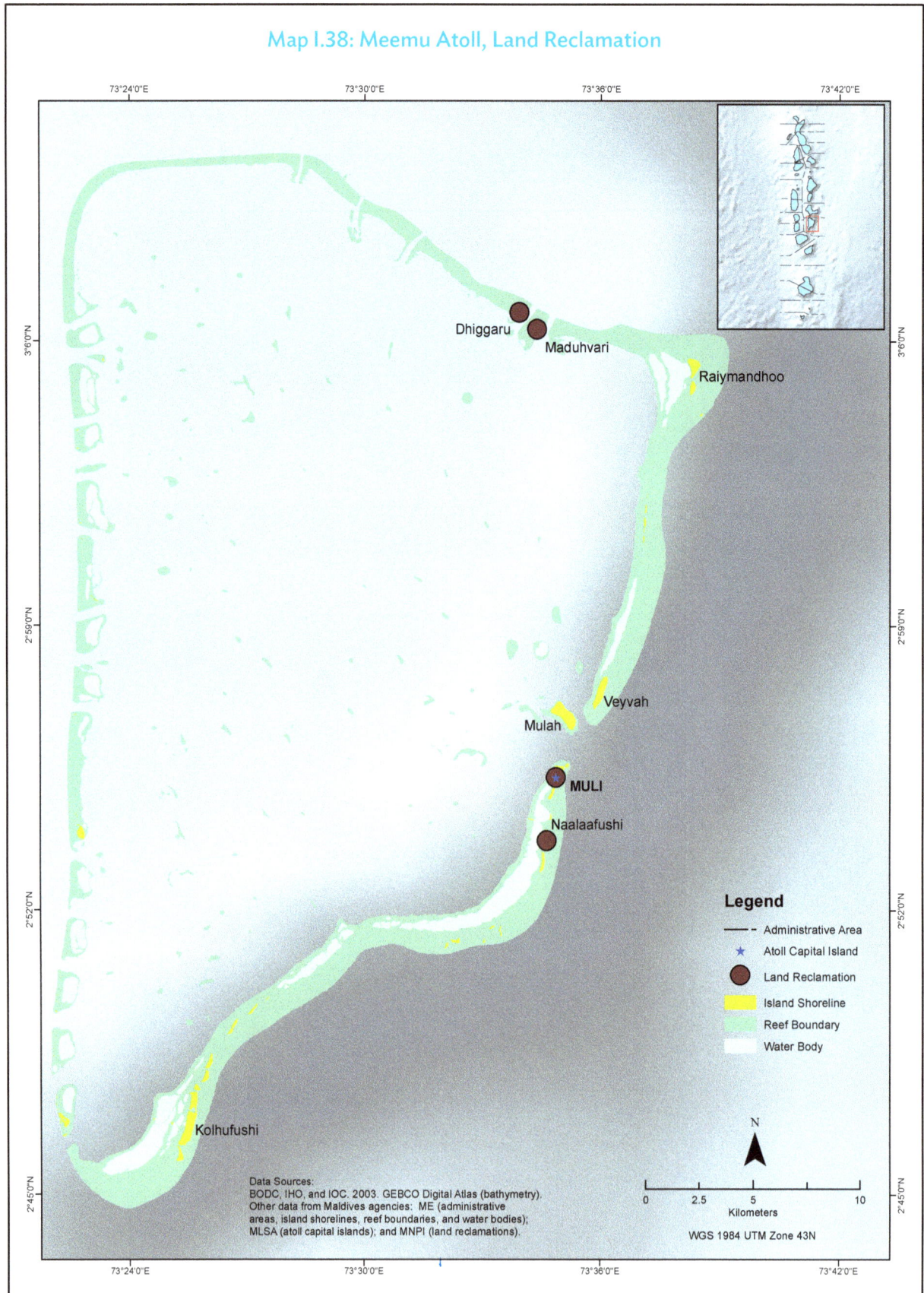

73°24'0"E   73°30'0"E   73°36'0"E   73°42'0"E

3°6'0"N

Dhiggaru

Maduhvari

Raiymandhoo

2°59'0"N

Veyvah

Mulah

MULI

Naalaafushi

2°52'0"N

Kolhufushi

2°45'0"N

Legend

— - — Administrative Area

★ Atoll Capital Island

● Land Reclamation

▮ Island Shoreline

▮ Reef Boundary

▮ Water Body

N

Data Sources:
BODC, IHO, and IOC. 2003. GEBCO Digital Atlas (bathymetry).
Other data from Maldives agencies: ME (administrative
areas, island shorelines, reef boundaries, and water bodies);
MLSA (atoll capital islands); and MNPI (land reclamations).

0   2.5   5   10
Kilometers

WGS 1984 UTM Zone 43N

Map I.39: Noonu Atoll, Land Reclamation

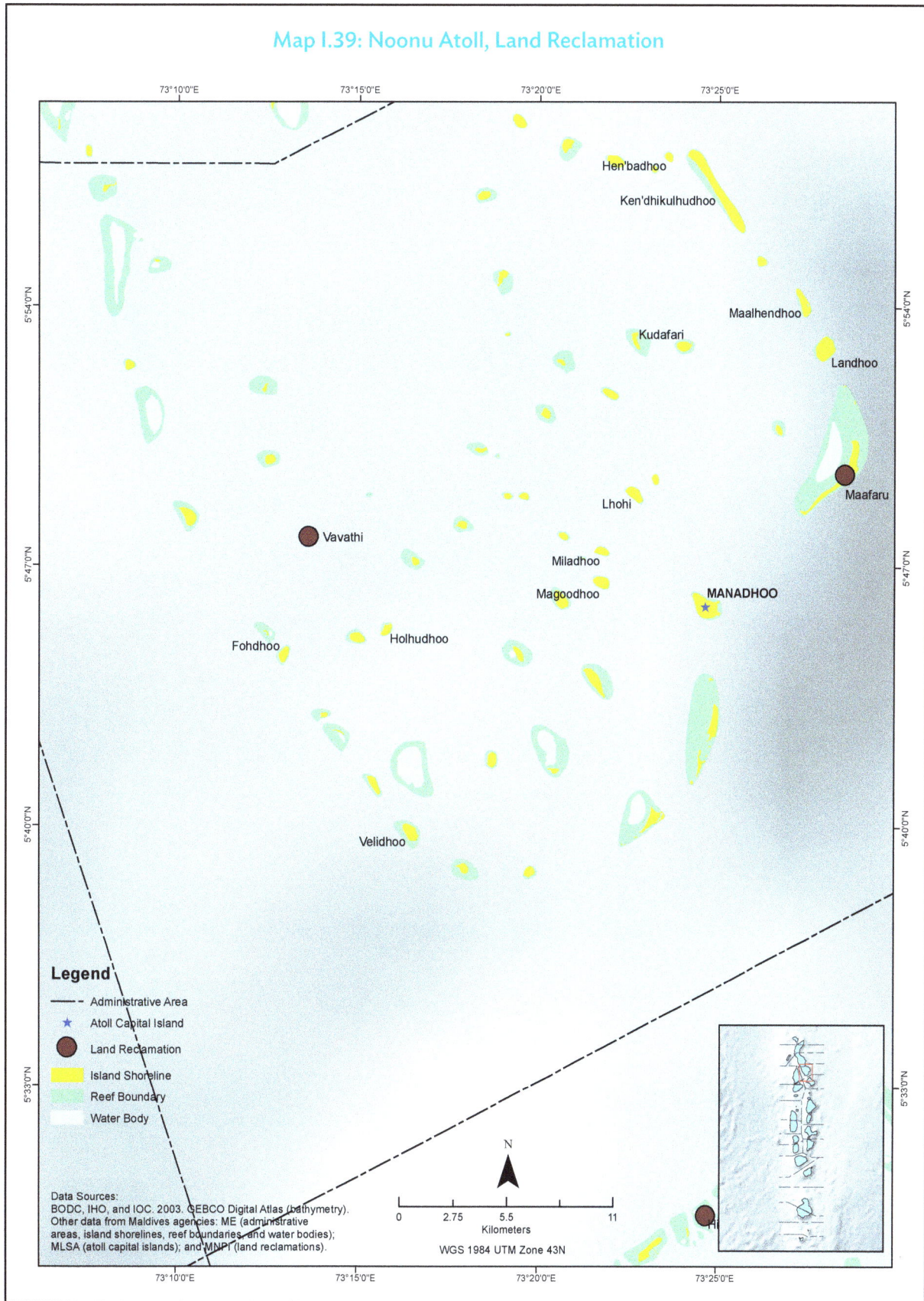

73°10'0"E 73°15'0"E 73°20'0"E 73°25'0"E

Hen'badhoo
Ken'dhikulhudhoo
Maalhendhoo
Kudafari
Landhoo
Maafaru
Lhohi
Vavathi
Miladhoo
Magoodhoo
MANADHOO
Fohdhoo
Holhudhoo
Velidhoo

5°54'0"N
5°47'0"N
5°40'0"N
5°33'0"N

Legend
- – – Administrative Area
- ★ Atoll Capital Island
- ● Land Reclamation
- Island Shoreline
- Reef Boundary
- Water Body

Data Sources:
BODC, IHO, and IOC. 2003. GEBCO Digital Atlas (bathymetry).
Other data from Maldives agencies: ME (administrative
areas, island shorelines, reef boundaries, and water bodies);
MLSA (atoll capital islands); and MNPI (land reclamations).

N

0   2.75   5.5   11
Kilometers
WGS 1984 UTM Zone 43N

Map I.40: North Malé Atoll, Land Reclamation

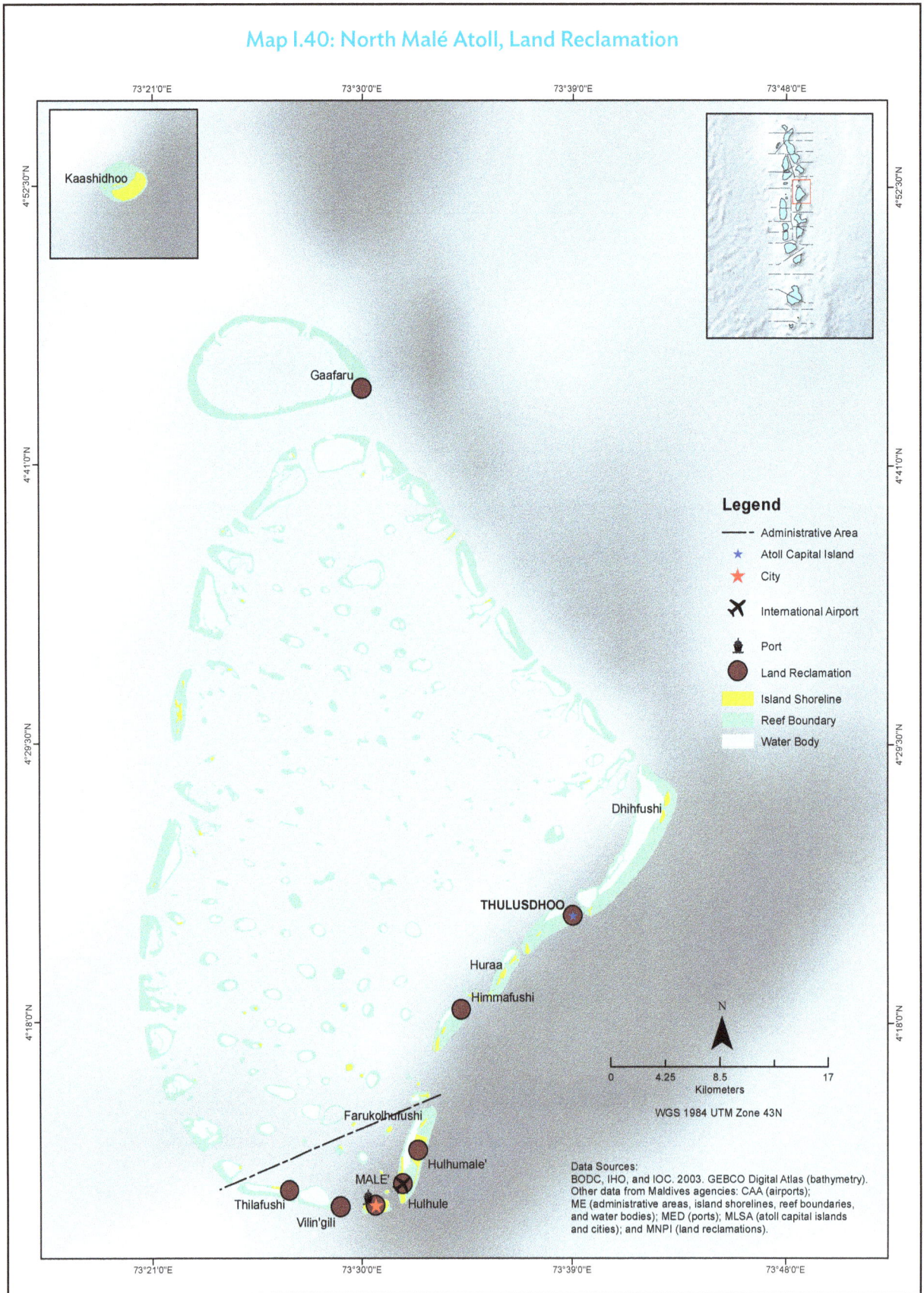

Kaashidhoo

Gaafaru

Dhihfushi

THULUSDHOO

Huraa

Himmafushi

Farukolhufushi

Hulhumale'

MALE'

Thilafushi

Vilin'gili

Hulhule

Legend

- - - Administrative Area
★ Atoll Capital Island
★ City
✈ International Airport
⚓ Port
● Land Reclamation
▮ Island Shoreline
▮ Reef Boundary
▮ Water Body

N

0   4.25   8.5   17
Kilometers

WGS 1984 UTM Zone 43N

Data Sources:
BODC, IHO, and IOC. 2003. GEBCO Digital Atlas (bathymetry).
Other data from Maldives agencies: CAA (airports);
ME (administrative areas, island shorelines, reef boundaries,
and water bodies); MED (ports); MLSA (atoll capital islands
and cities); and MNPI (land reclamations).

Map I.41: Shaviyani Atoll, Land Reclamation

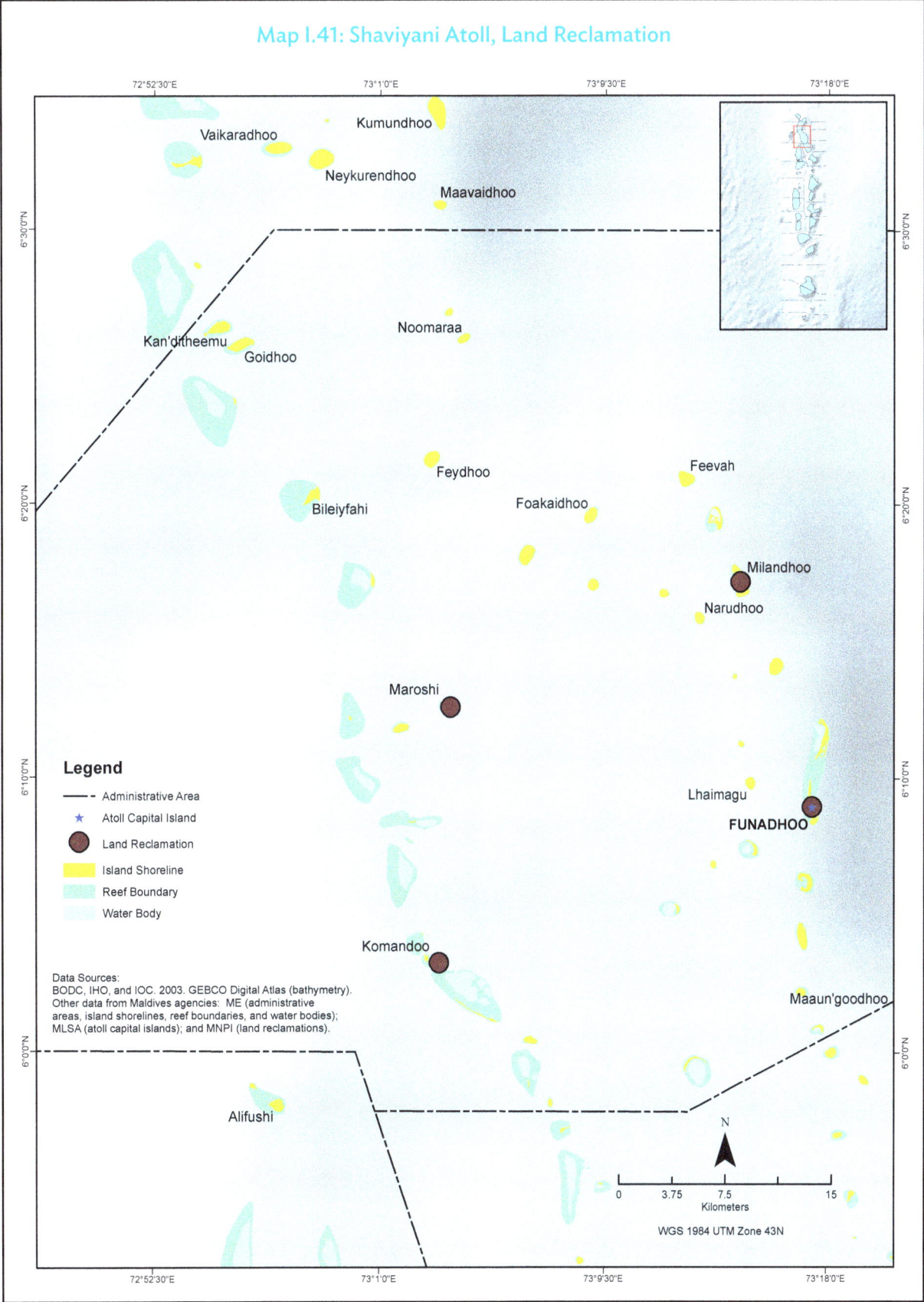

72°52'30"E   73°1'0"E   73°9'30"E   73°18'0"E

Kumundhoo

Vaikaradhoo

Neykurendhoo

Maavaidhoo

6°30'0"N

Noomaraa

Kan'ditheemu

Goidhoo

Feevah

6°20'0"N

Feydhoo

Bileiyfahi

Foakaidhoo

Milandhoo

Narudhoo

Maroshi

Lhaimagu

6°10'0"N

FUNADHOO

Legend

— — Administrative Area

★ Atoll Capital Island

⬤ Land Reclamation

▮ Island Shoreline

▮ Reef Boundary

▮ Water Body

Data Sources:
BODC, IHO, and IOC. 2003. GEBCO Digital Atlas (bathymetry).
Other data from Maldives agencies: ME (administrative
areas, island shorelines, reef boundaries, and water bodies);
MLSA (atoll capital islands); and MNPI (land reclamations).

Komandoo

Maaun'goodhoo

6°0'0"N

Alifushi

N

0   3.75   7.5   15
Kilometers

WGS 1984 UTM Zone 43N

72°52'30"E   73°1'0"E   73°9'30"E   73°18'0"E

Map I.42: South Malé Atoll, Land Reclamation

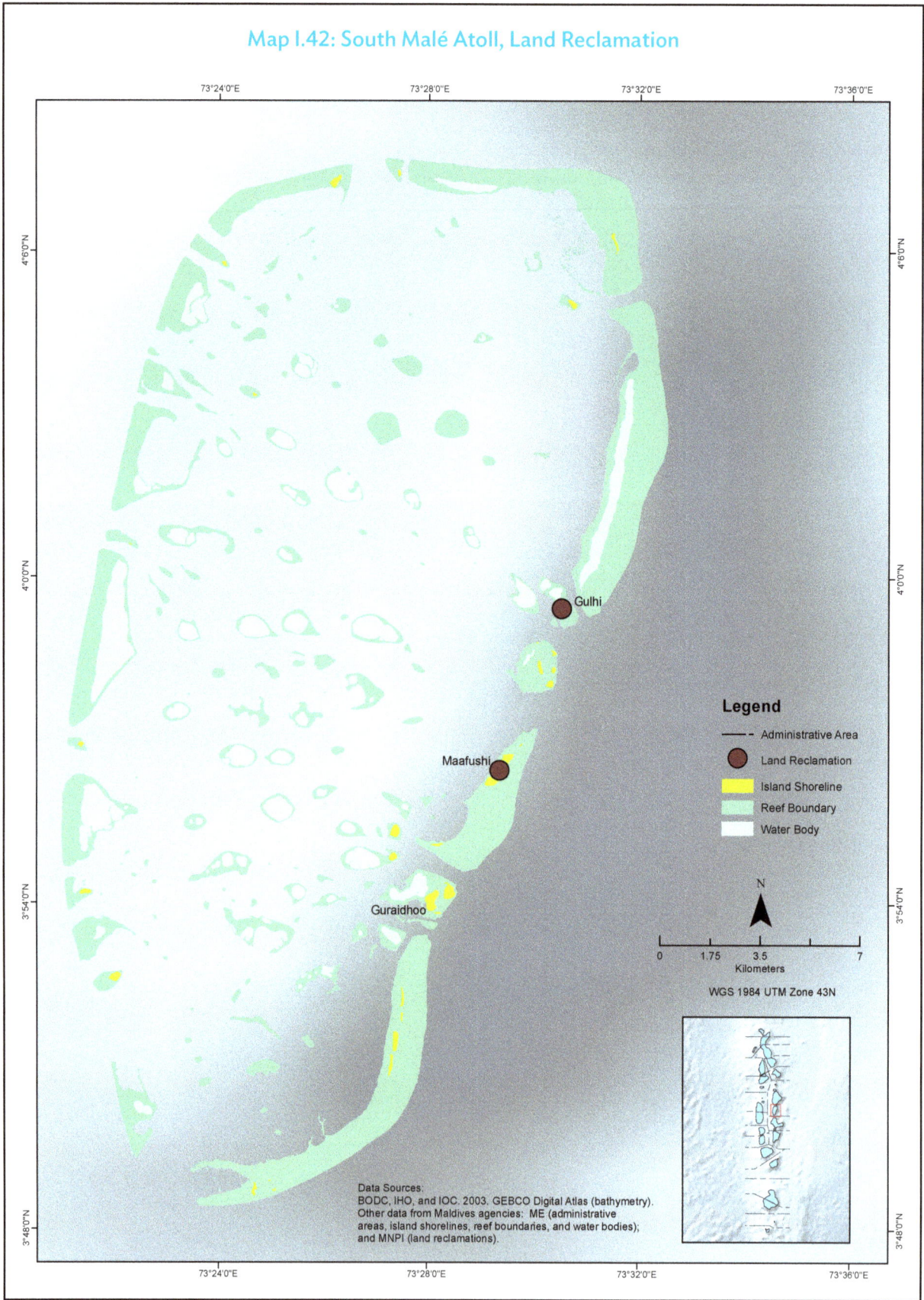

73°24'0"E 73°28'0"E 73°32'0"E 73°36'0"E

4°6'0"N
4°0'0"N
3°54'0"N
3°48'0"N

Gulhi

Maafushi

Guraidhoo

Legend

- - - Administrative Area
⬤ Land Reclamation
▮ Island Shoreline
▮ Reef Boundary
▯ Water Body

N

0   1.75   3.5   7
Kilometers

WGS 1984 UTM Zone 43N

Data Sources:
BODC, IHO, and IOC. 2003. GEBCO Digital Atlas (bathymetry).
Other data from Maldives agencies: ME (administrative
areas, island shorelines, reef boundaries, and water bodies);
and MNPI (land reclamations).

Map I.43: Thaa Atoll, Land Reclamation

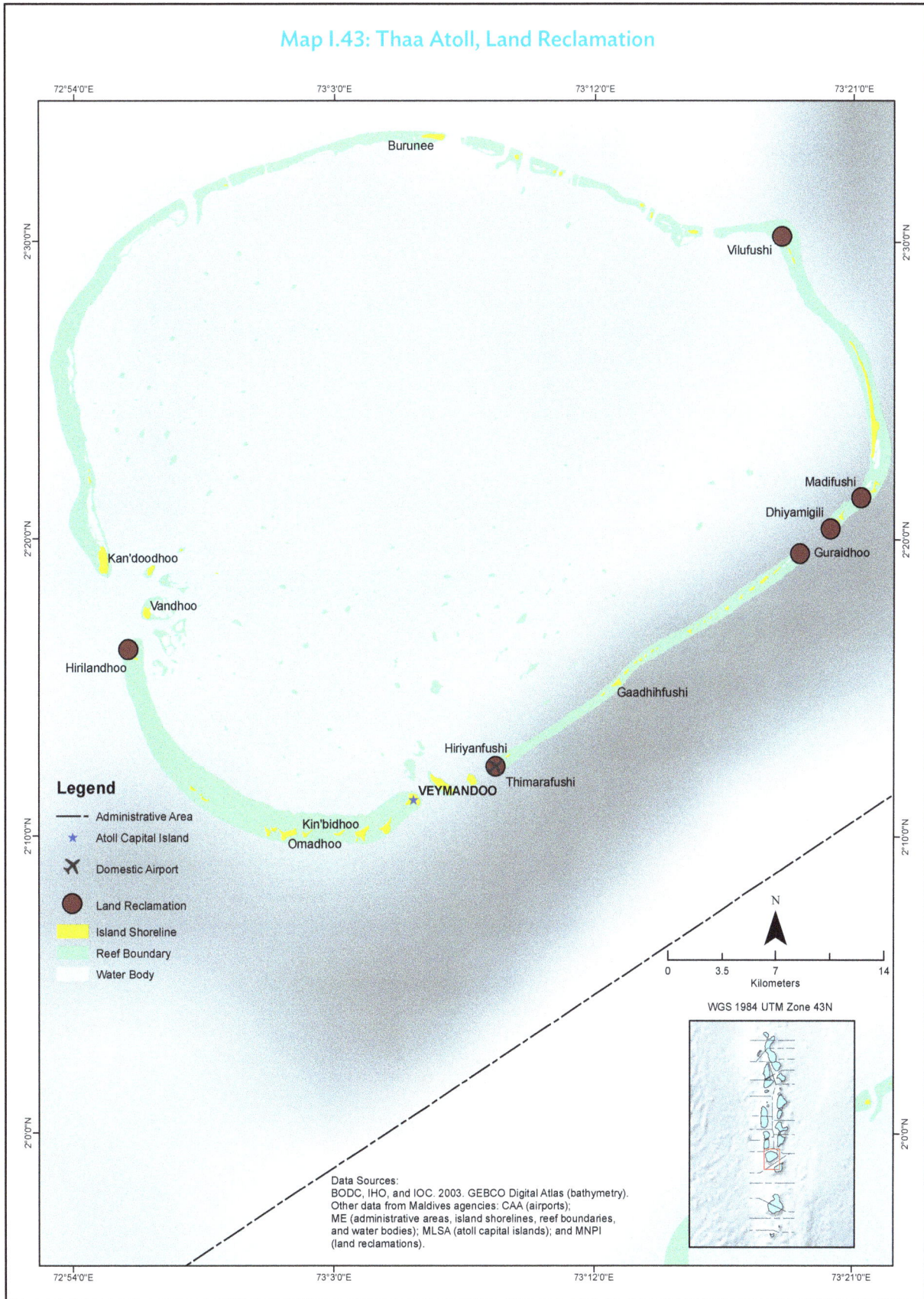

Legend

- Administrative Area
- Atoll Capital Island
- Domestic Airport
- Land Reclamation
- Island Shoreline
- Reef Boundary
- Water Body

Burunee

Vilufushi

Madifushi

Dhiyamigili

Guraidhoo

Kan'doodhoo

Vandhoo

Hirilandhoo

Gaadhihfushi

Hiriyanfushi

VEYMANDOO

Thimarafushi

Kin'bidhoo

Omadhoo

N

0   3.5   7   14
Kilometers

WGS 1984 UTM Zone 43N

Data Sources:
BODC, IHO, and IOC. 2003. GEBCO Digital Atlas (bathymetry).
Other data from Maldives agencies: CAA (airports);
ME (administrative areas, island shorelines, reef boundaries,
and water bodies); MLSA (atoll capital islands); and MNPI
(land reclamations).

Land Management

Land space is a necessary resource for development. Maldives has only less than 250 square kilometers (km²) of land. With its limited available land, allocation of space for specific land use is a challenge. Currently, the country has more vegetated land cover (shrubs, herbs, forest, and palm trees) than built-up areas (high density urban areas, road, airport, and low density urban areas). Huge land is allocated for beaches (20.9 km²) and island resorts (16.8 km²). A small portion (6.7 km²) is for agriculture. Other spaces are classified as inland water and wetlands (Table I.2).

Table I.2: Land Use and Land Cover, Maldives

Land Use and Land Cover	Area (km²)	Percentage
Land Use		
High density urban areas	1.79	0.76
Roads	4.66	1.98
Airports	5.80	2.46
Harbors	7.32	3.11
Island resorts	16.78	7.12
Low density urban areas	29.62	12.56
Agricultural areas	6.65	2.82
Land Cover		
Inland waters	2.97	1.26
Wetlands	3.39	1.44
Barren, sparsely vegetated areas	8.88	3.77
Beaches and sand	20.85	8.84
Shrubs and/or herbaceous vegetation areas	36.07	15.30
Forests	40.86	17.33
Palm trees	50.10	21.25
Total	**235.74**	**100.00**

km² = square kilometer.

Note: Total may not add up due to rounding.

Source: Government of Maldives, Ministry of Environment, 2016.

Land cover. Maldives' national tree, the palm tree, covers a fifth of the country's land area. Together with palm trees, forests and shrublands cover more than half of the country's land (photo by Shifaaz Shamoon).

Buildings stand next to one another along the narrow roads of Malé City in Maldives. Less than 1% of Maldives has high density urban land use classification (photo by Shahee Ilyas).

Map I.44: Addu City, Land Use and Land Cover

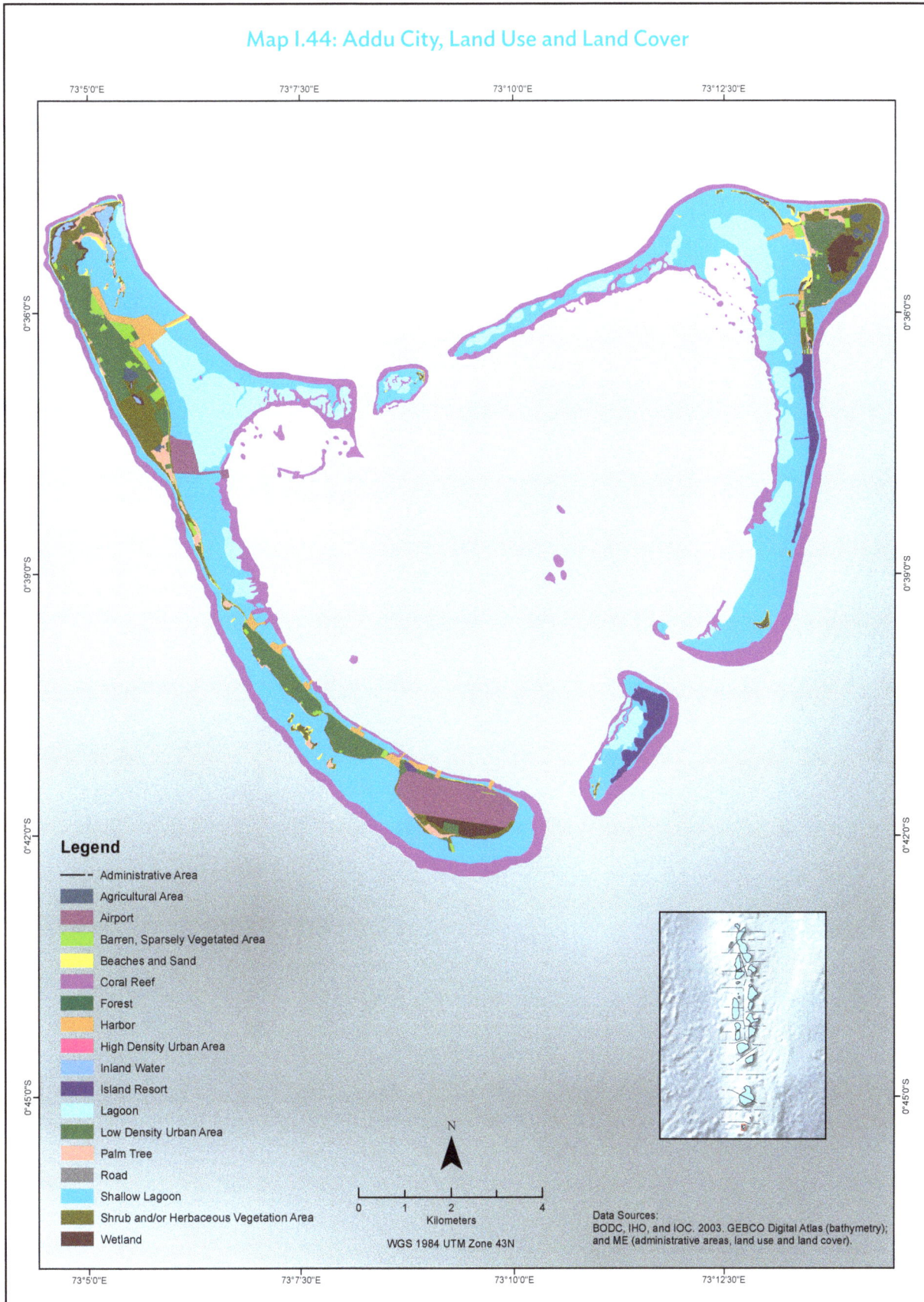

73°5'0"E 73°7'30"E 73°10'0"E 73°12'30"E

0°36'0"S

0°39'0"S

0°42'0"S

0°45'0"S

Legend

— Administrative Area
▮ Agricultural Area
▮ Airport
▮ Barren, Sparsely Vegetated Area
▮ Beaches and Sand
▮ Coral Reef
▮ Forest
▮ Harbor
▮ High Density Urban Area
▮ Inland Water
▮ Island Resort
▮ Lagoon
▮ Low Density Urban Area
▮ Palm Tree
▮ Road
▮ Shallow Lagoon
▮ Shrub and/or Herbaceous Vegetation Area
▮ Wetland

N

0 1 2 4
Kilometers

WGS 1984 UTM Zone 43N

Data Sources:
BODC, IHO, and IOC. 2003. GEBCO Digital Atlas (bathymetry);
and ME (administrative areas, land use and land cover).

73°5'0"E 73°7'30"E 73°10'0"E 73°12'30"E

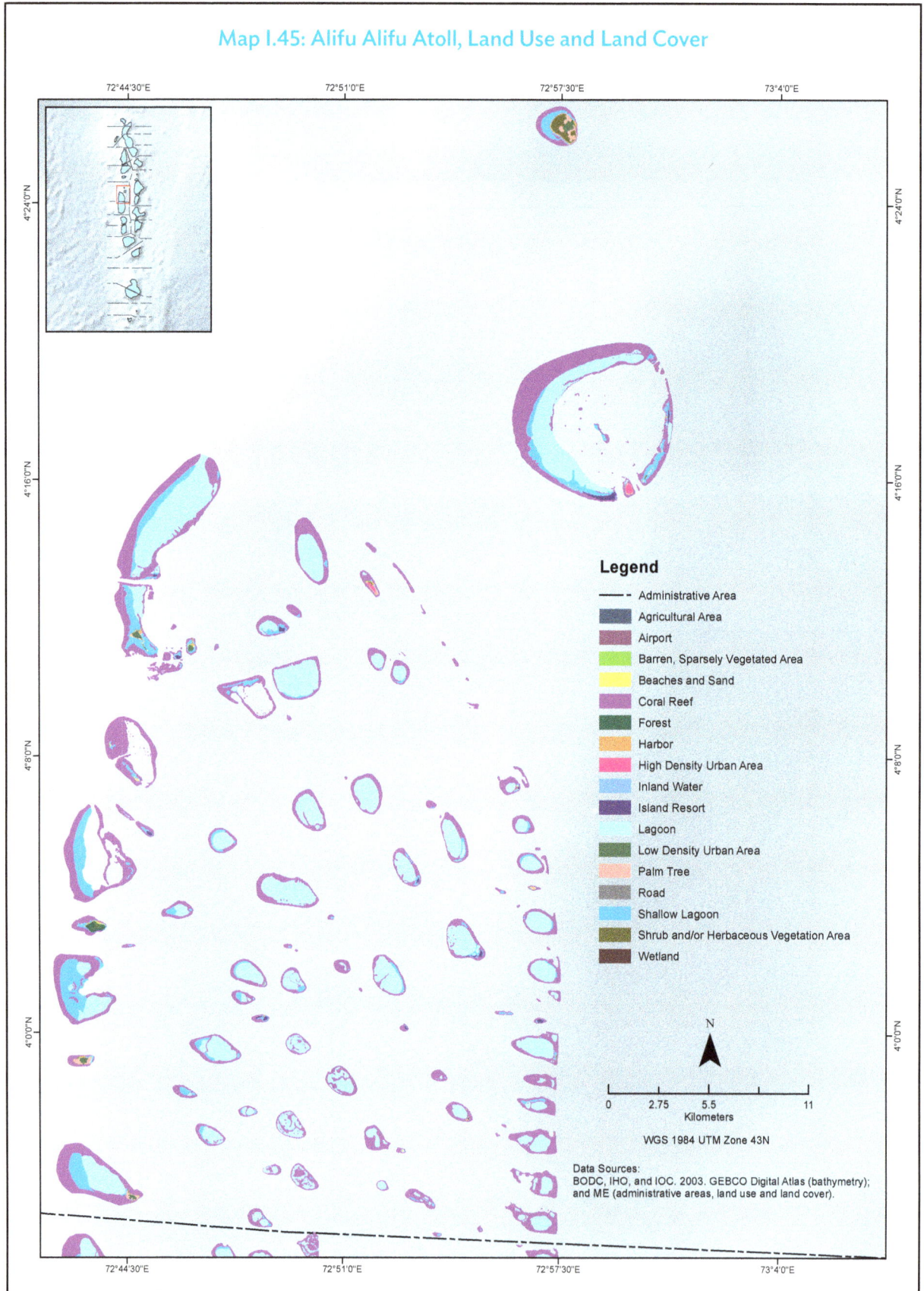

Map I.45: Alifu Alifu Atoll, Land Use and Land Cover

Legend

— — Administrative Area
Agricultural Area
Airport
Barren, Sparsely Vegetated Area
Beaches and Sand
Coral Reef
Forest
Harbor
High Density Urban Area
Inland Water
Island Resort
Lagoon
Low Density Urban Area
Palm Tree
Road
Shallow Lagoon
Shrub and/or Herbaceous Vegetation Area
Wetland

N

0   2.75   5.5   11
Kilometers

WGS 1984 UTM Zone 43N

Data Sources:
BODC, IHO, and IOC. 2003. GEBCO Digital Atlas (bathymetry);
and ME (administrative areas, land use and land cover).

Map I.46: Alifu Dhaalu Atoll, Land Use and Land Cover

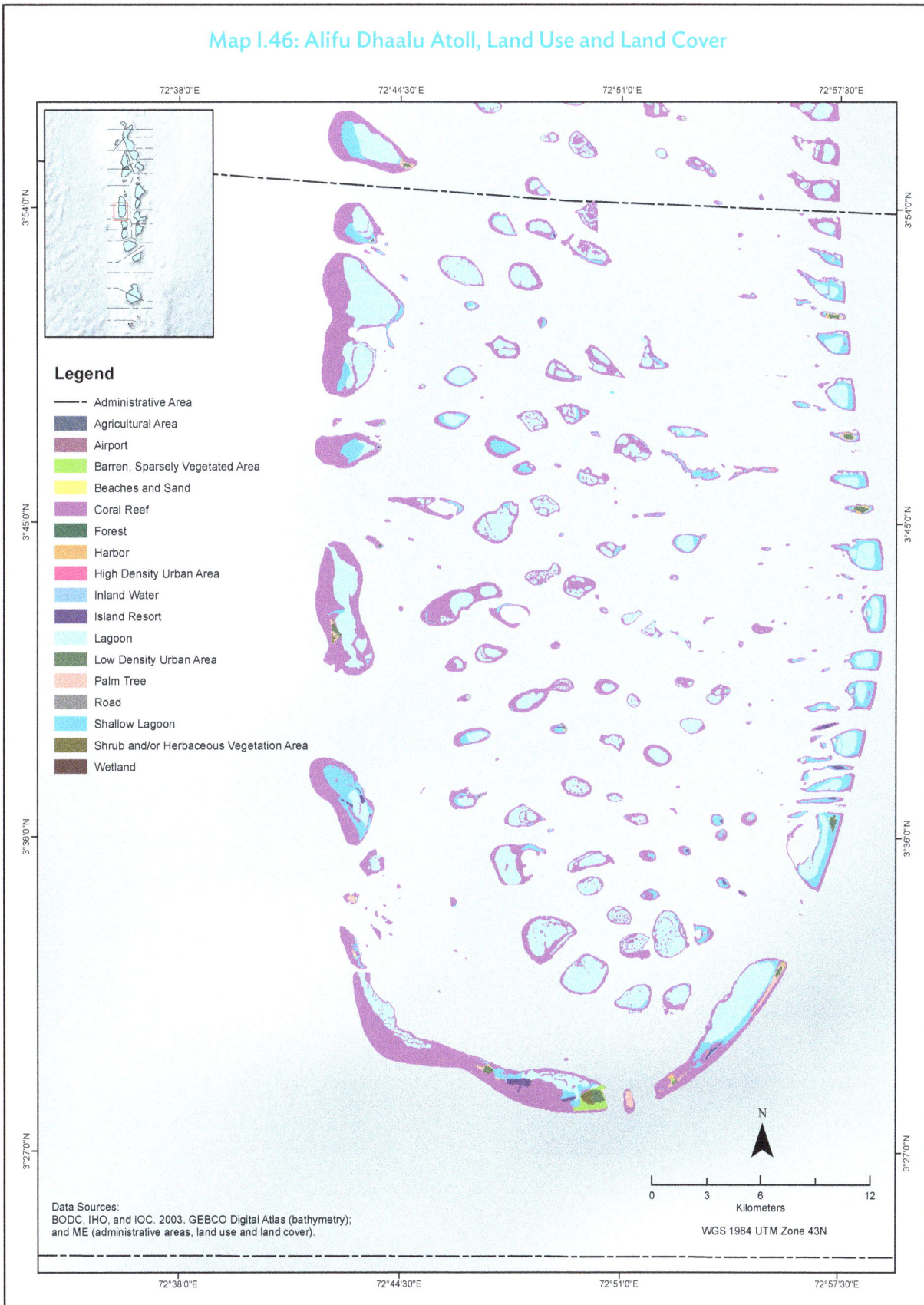

Legend

- — Administrative Area
- Agricultural Area
- Airport
- Barren, Sparsely Vegetated Area
- Beaches and Sand
- Coral Reef
- Forest
- Harbor
- High Density Urban Area
- Inland Water
- Island Resort
- Lagoon
- Low Density Urban Area
- Palm Tree
- Road
- Shallow Lagoon
- Shrub and/or Herbaceous Vegetation Area
- Wetland

Data Sources:
BODC, IHO, and IOC. 2003. GEBCO Digital Atlas (bathymetry);
and ME (administrative areas, land use and land cover).

WGS 1984 UTM Zone 43N

0 3 6 12
Kilometers

Map I.47: Baa Atoll, Land Use and Land Cover

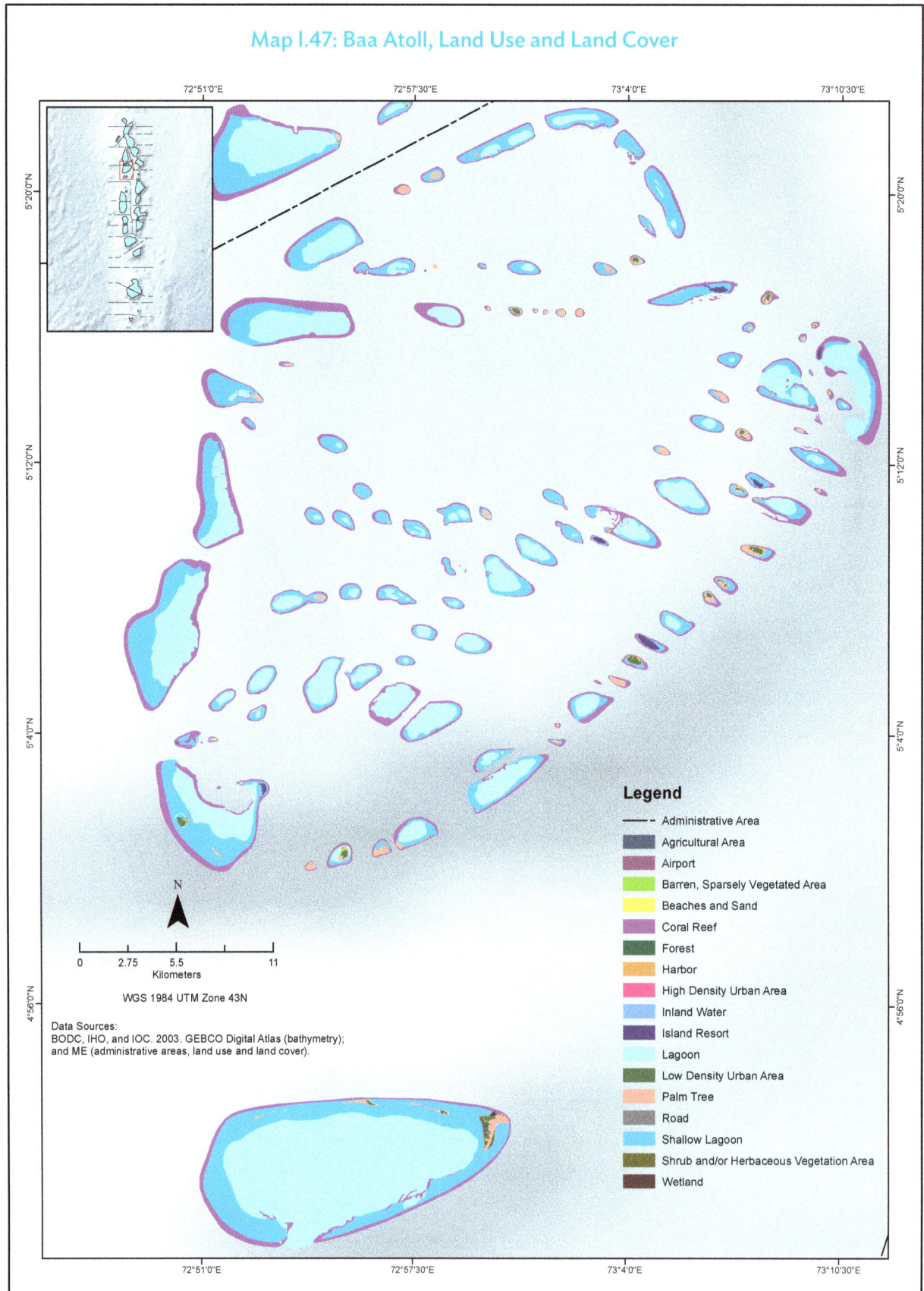

Legend

- Administrative Area
- Agricultural Area
- Airport
- Barren, Sparsely Vegetated Area
- Beaches and Sand
- Coral Reef
- Forest
- Harbor
- High Density Urban Area
- Inland Water
- Island Resort
- Lagoon
- Low Density Urban Area
- Palm Tree
- Road
- Shallow Lagoon
- Shrub and/or Herbaceous Vegetation Area
- Wetland

N

0   2.75   5.5   11
Kilometers

WGS 1984 UTM Zone 43N

Data Sources:
BODC, IHO, and IOC. 2003. GEBCO Digital Atlas (bathymetry);
and ME (administrative areas, land use and land cover).

Map I.48: Dhaalu Atoll, Land Use and Land Cover

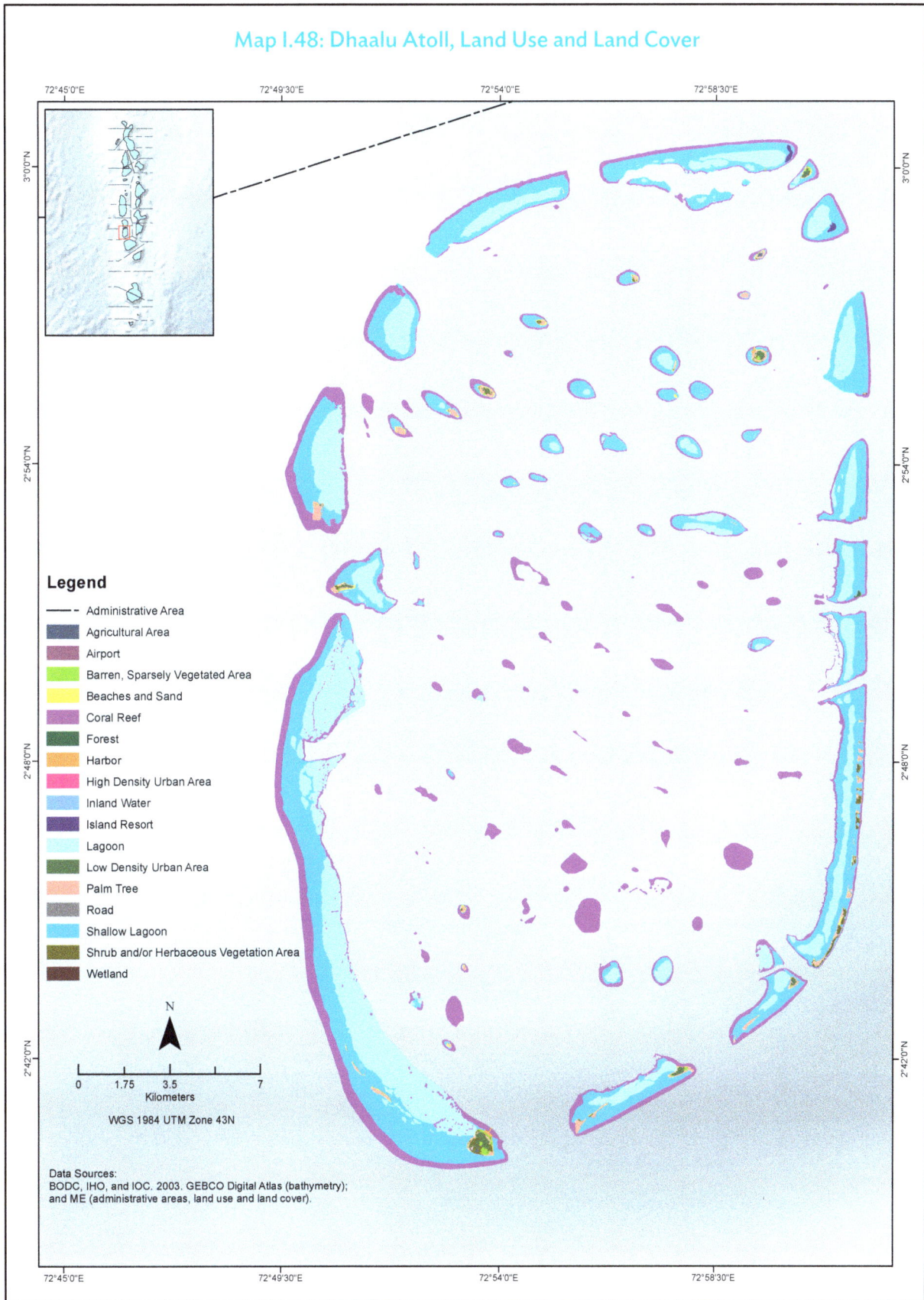

72°45'0"E   72°49'30"E   72°54'0"E   72°58'30"E

3°0'0"N

2°54'0"N

2°48'0"N

2°42'0"N

Legend

- — — Administrative Area
- Agricultural Area
- Airport
- Barren, Sparsely Vegetated Area
- Beaches and Sand
- Coral Reef
- Forest
- Harbor
- High Density Urban Area
- Inland Water
- Island Resort
- Lagoon
- Low Density Urban Area
- Palm Tree
- Road
- Shallow Lagoon
- Shrub and/or Herbaceous Vegetation Area
- Wetland

N

0   1.75   3.5   7
Kilometers

WGS 1984 UTM Zone 43N

Data Sources:
BODC, IHO, and IOC. 2003. GEBCO Digital Atlas (bathymetry);
and ME (administrative areas, land use and land cover).

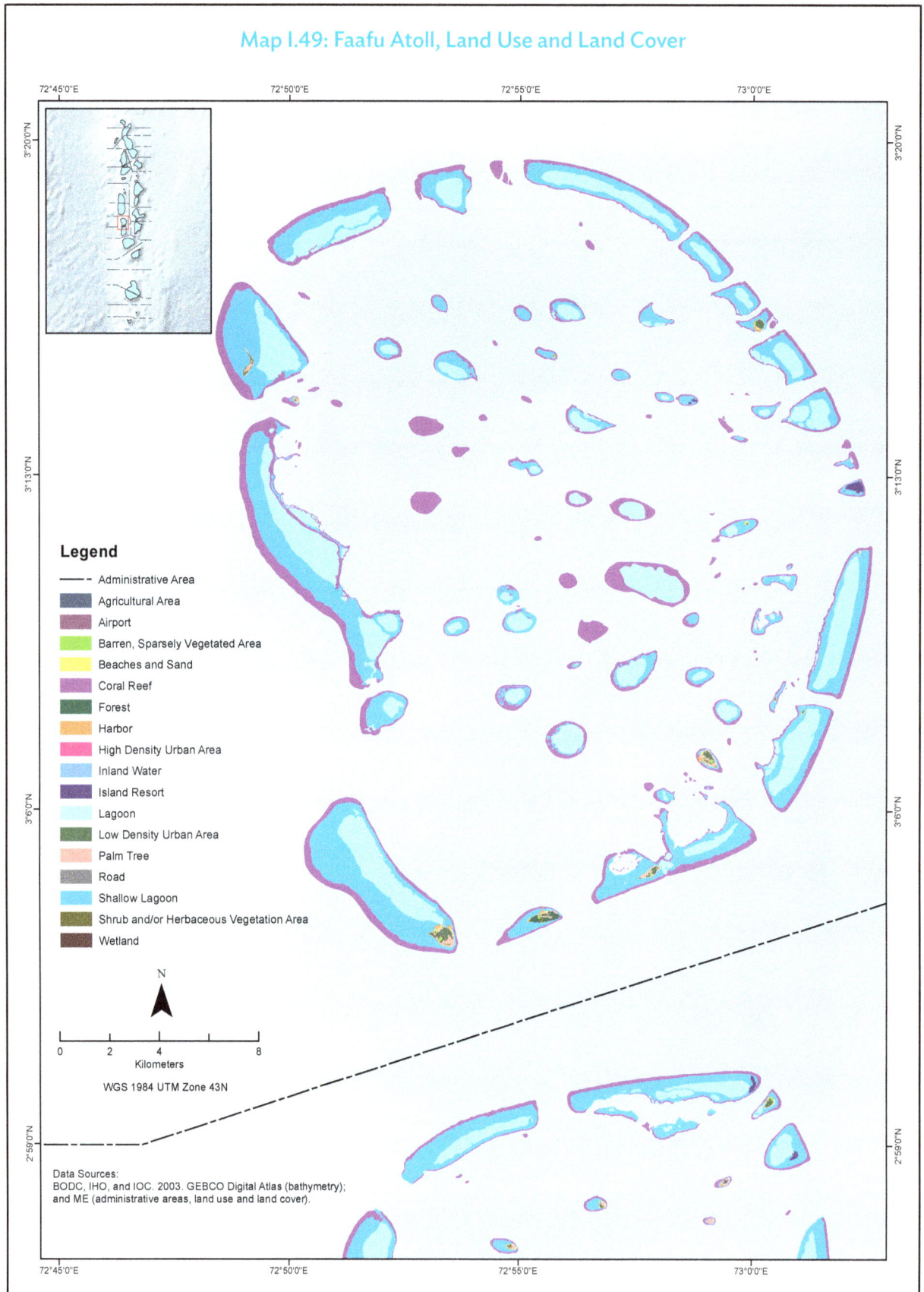

Map I.49: Faafu Atoll, Land Use and Land Cover

72°45'0"E   72°50'0"E   72°55'0"E   73°0'0"E

3°20'0"N

3°13'0"N

3°6'0"N

2°59'0"N

Legend

— - — Administrative Area

Agricultural Area

Airport

Barren, Sparsely Vegetated Area

Beaches and Sand

Coral Reef

Forest

Harbor

High Density Urban Area

Inland Water

Island Resort

Lagoon

Low Density Urban Area

Palm Tree

Road

Shallow Lagoon

Shrub and/or Herbaceous Vegetation Area

Wetland

N

0   2   4   8
Kilometers

WGS 1984 UTM Zone 43N

Data Sources:
BODC, IHO, and IOC. 2003. GEBCO Digital Atlas (bathymetry);
and ME (administrative areas, land use and land cover).

Map I.50: Gaafu Alifu Atoll, Land Use and Land Cover

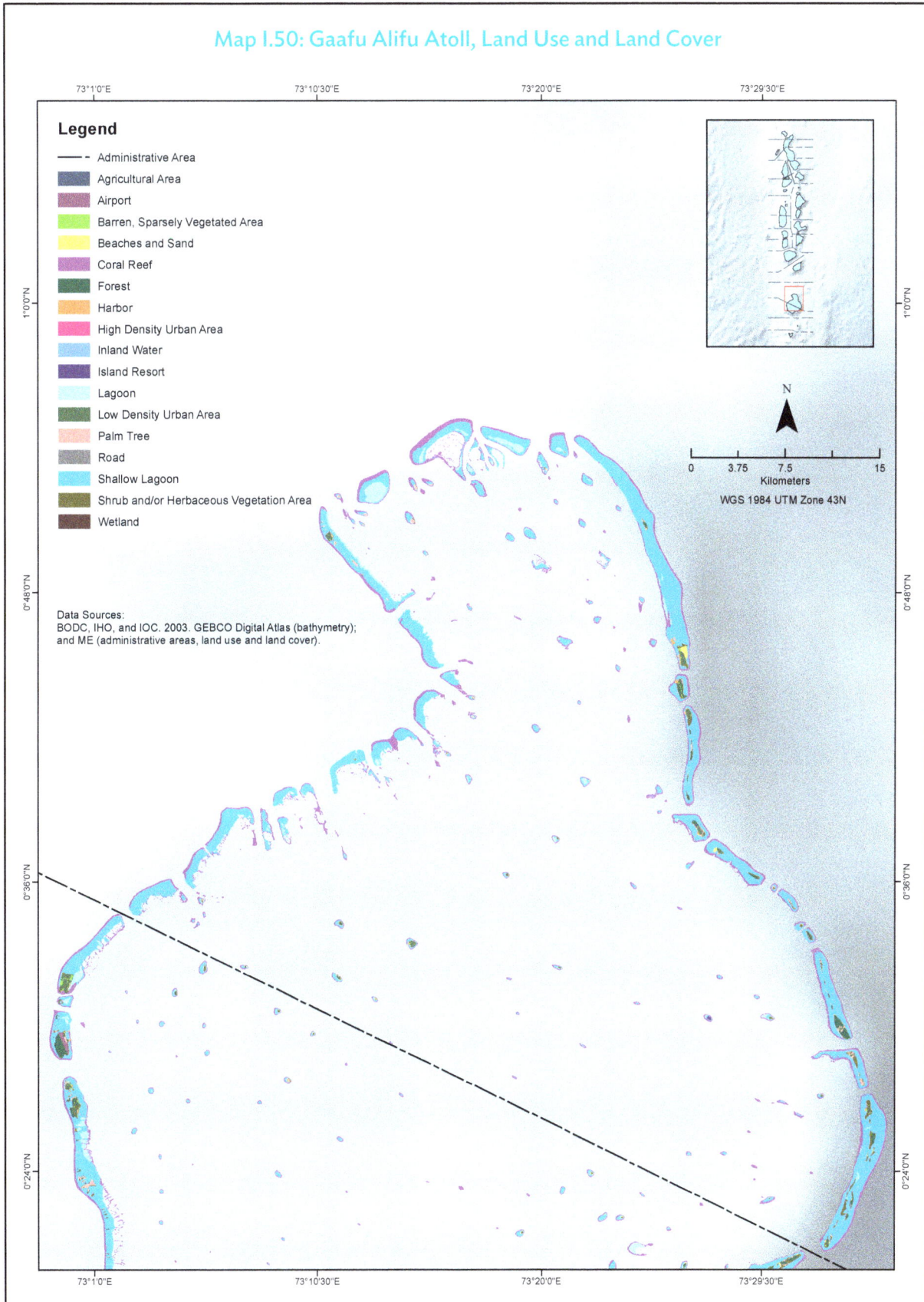

73°1'0"E   73°10'30"E   73°20'0"E   73°29'30"E

Legend

— - — Administrative Area

▬ Agricultural Area

▬ Airport

▬ Barren, Sparsely Vegetated Area

▬ Beaches and Sand

▬ Coral Reef

▬ Forest

▬ Harbor

▬ High Density Urban Area

▬ Inland Water

▬ Island Resort

▬ Lagoon

▬ Low Density Urban Area

▬ Palm Tree

▬ Road

▬ Shallow Lagoon

▬ Shrub and/or Herbaceous Vegetation Area

▬ Wetland

Data Sources:
BODC, IHO, and IOC. 2003. GEBCO Digital Atlas (bathymetry);
and ME (administrative areas, land use and land cover).

N

0   3.75   7.5   15
        Kilometers
WGS 1984 UTM Zone 43N

1°0'0"N

0°48'0"N

0°36'0"N

0°24'0"N

Map I.51: Gaafu Dhaalu Atoll, Land Use and Land Cover

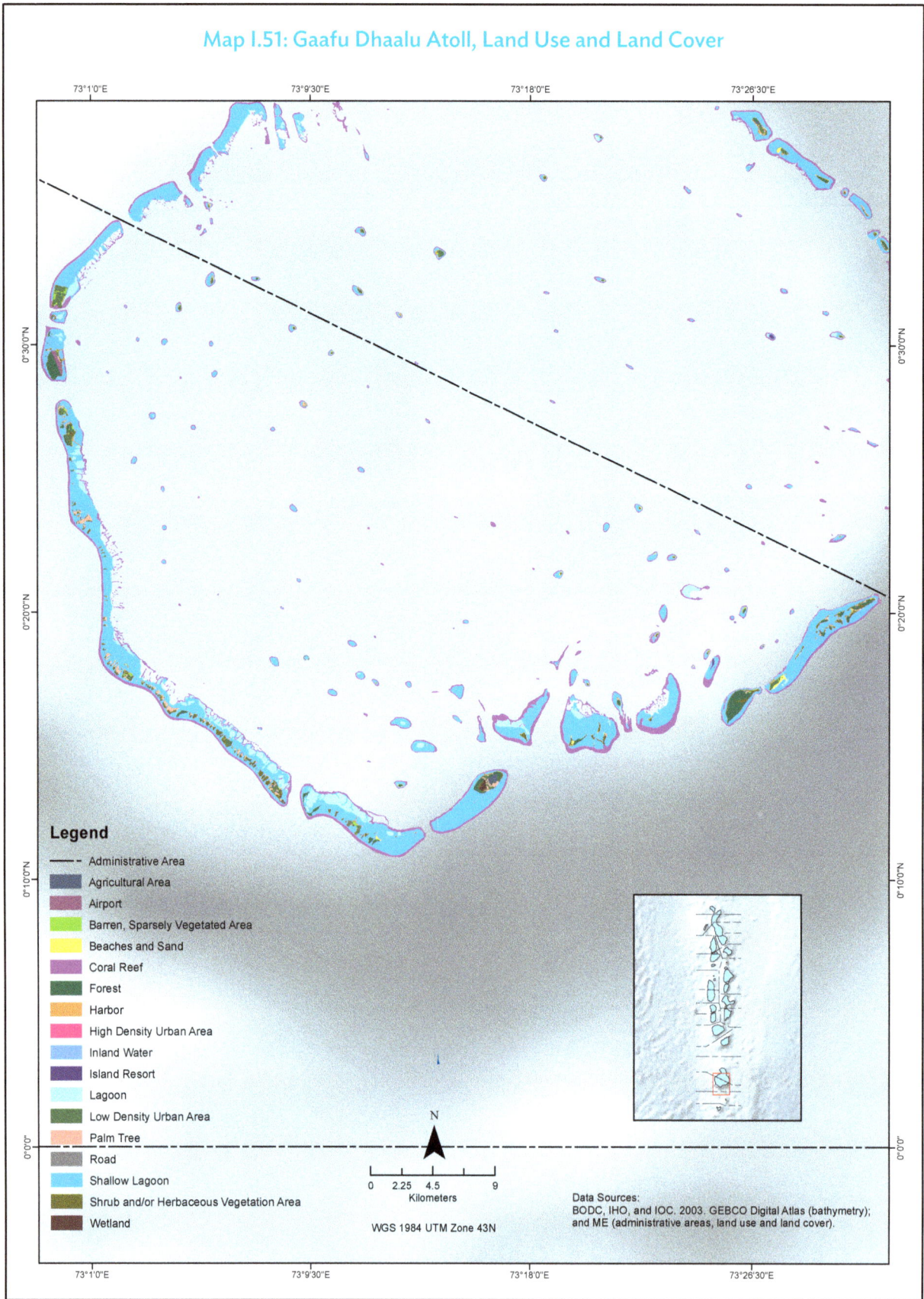

73°1'0"E   73°9'30"E   73°18'0"E   73°26'30"E

Legend

— Administrative Area
▨ Agricultural Area
▨ Airport
▨ Barren, Sparsely Vegetated Area
▨ Beaches and Sand
▨ Coral Reef
▨ Forest
▨ Harbor
▨ High Density Urban Area
▨ Inland Water
▨ Island Resort
▨ Lagoon
▨ Low Density Urban Area
▨ Palm Tree
▨ Road
▨ Shallow Lagoon
▨ Shrub and/or Herbaceous Vegetation Area
▨ Wetland

N

0   2.25   4.5   9
Kilometers

WGS 1984 UTM Zone 43N

Data Sources:
BODC, IHO, and IOC. 2003. GEBCO Digital Atlas (bathymetry);
and ME (administrative areas, land use and land cover).

73°1'0"E   73°9'30"E   73°18'0"E   73°26'30"E

Map I.52: Gnaviyani Atoll, Land Use and Land Cover

73°24'35"E   73°25'10"E   73°25'45"E   73°26'20"E

0°16'40"S

0°17'30"S

0°18'20"S

0°19'10"S

Legend

- – – Administrative Area
- Agricultural Area
- Airport
- Barren, Sparsely Vegetated Area
- Beaches and Sand
- Coral Reef
- Forest
- Harbor
- High Density Urban Area
- Inland Water
- Island Resort
- Lagoon
- Low Density Urban Area
- Palm Tree
- Road
- Shallow Lagoon
- Shrub and/or Herbaceous Vegetation Area
- Wetland

N

0   0.25   0.5   1
Kilometers

WGS 1984 UTM Zone 43N

Data Sources:
BODC, IHO, and IOC. 2003. GEBCO Digital Atlas (bathymetry);
and ME (administrative areas, land use and land cover).

73°24'35"E   73°25'10"E   73°25'45"E   73°26'20"E

Map I.53: Haa Alifu Atoll, Land Use and Land Cover

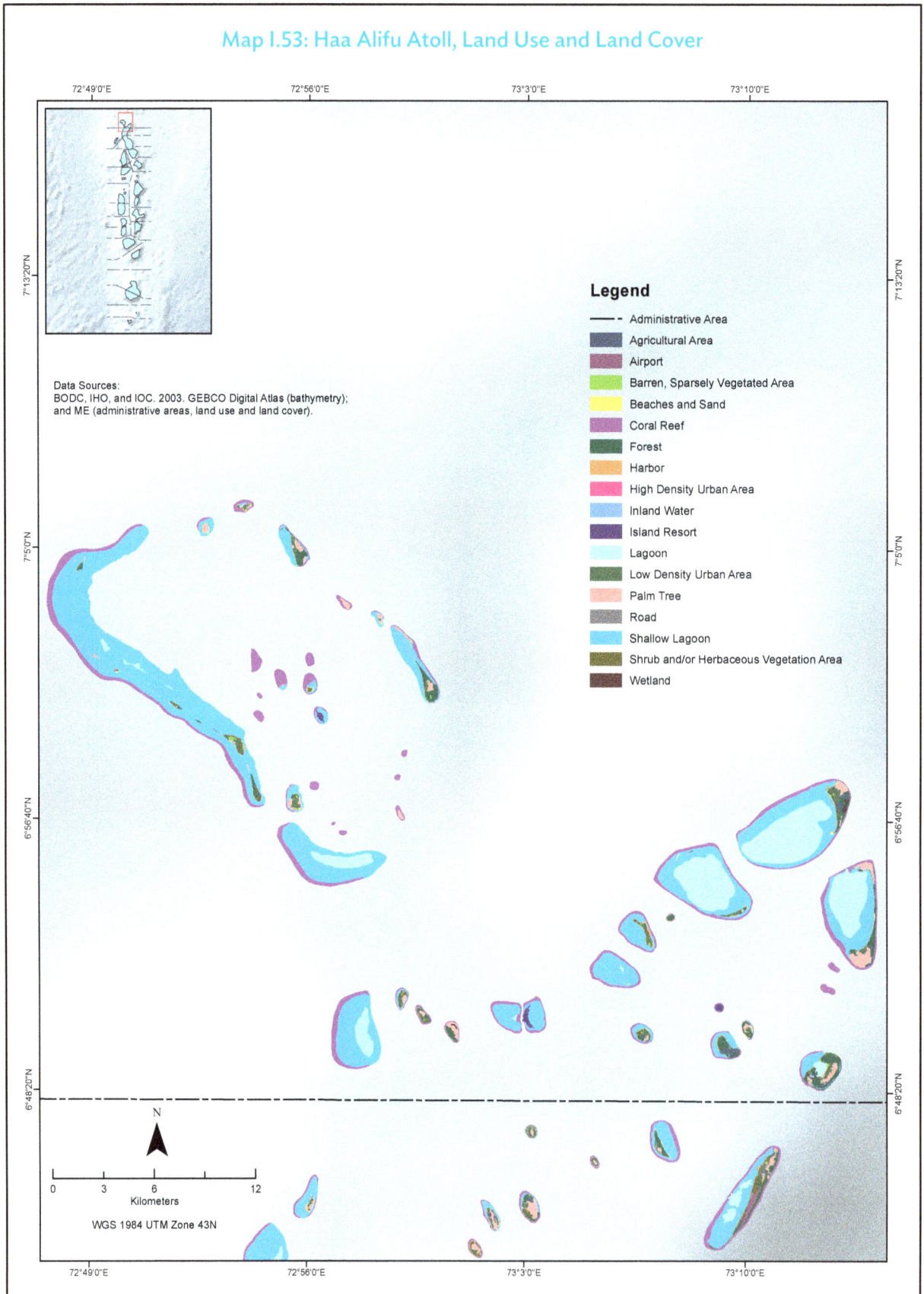

72°49'0"E   72°56'0"E   73°3'0"E   73°10'0"E

7°13'20"N

Data Sources:
BODC, IHO, and IOC. 2003. GEBCO Digital Atlas (bathymetry);
and ME (administrative areas, land use and land cover).

7°5'0"N

6°56'40"N

6°48'20"N

Legend

- – – Administrative Area
- Agricultural Area
- Airport
- Barren, Sparsely Vegetated Area
- Beaches and Sand
- Coral Reef
- Forest
- Harbor
- High Density Urban Area
- Inland Water
- Island Resort
- Lagoon
- Low Density Urban Area
- Palm Tree
- Road
- Shallow Lagoon
- Shrub and/or Herbaceous Vegetation Area
- Wetland

N

0   3   6   12
Kilometers

WGS 1984 UTM Zone 43N

Map I.54: Haa Dhaalu Atoll, Land Use and Land Cover

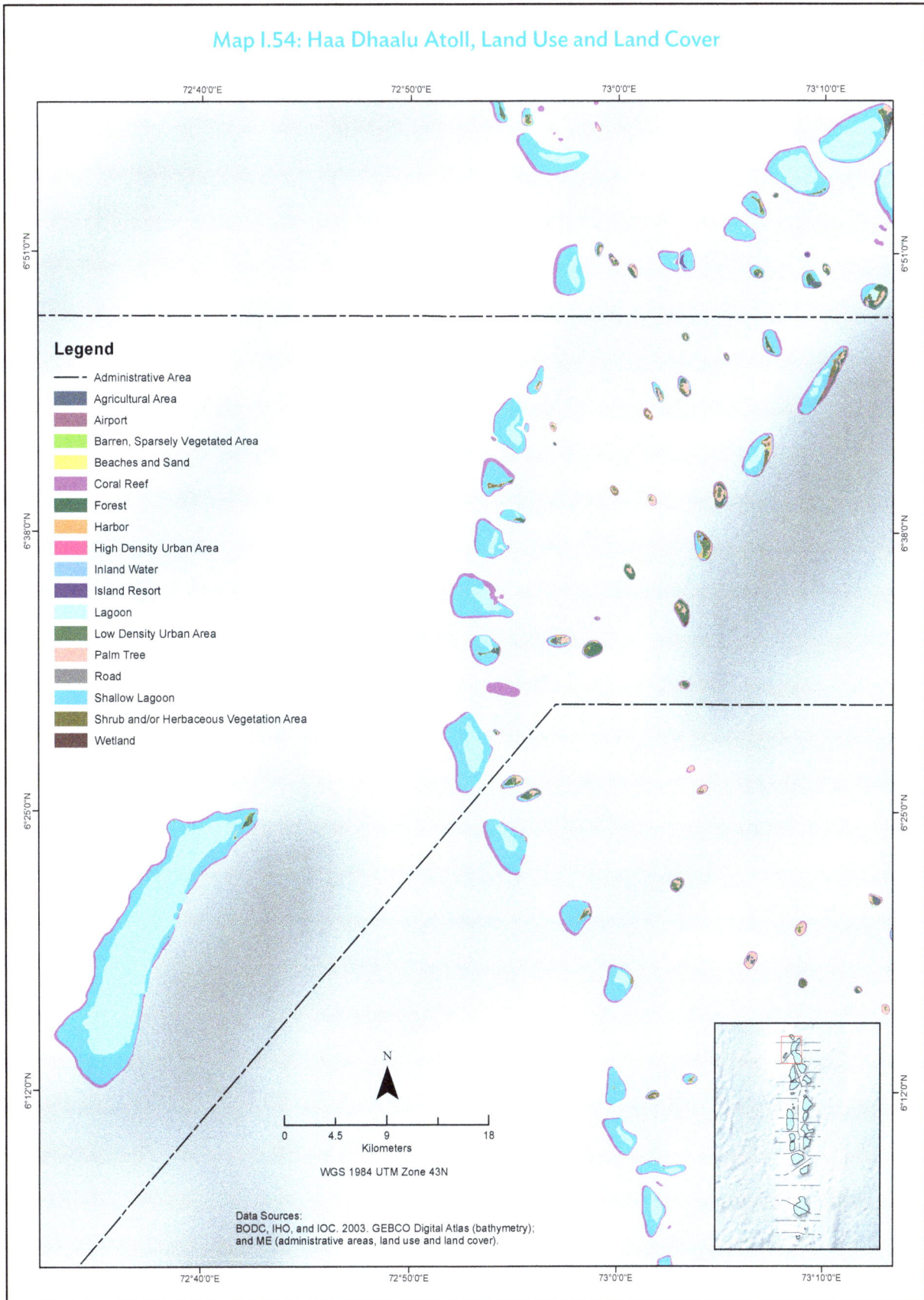

72°40'0"E   72°50'0"E   73°0'0"E   73°10'0"E

6°51'0"N

6°38'0"N

6°25'0"N

6°12'0"N

Legend

- – – – Administrative Area
- Agricultural Area
- Airport
- Barren, Sparsely Vegetated Area
- Beaches and Sand
- Coral Reef
- Forest
- Harbor
- High Density Urban Area
- Inland Water
- Island Resort
- Lagoon
- Low Density Urban Area
- Palm Tree
- Road
- Shallow Lagoon
- Shrub and/or Herbaceous Vegetation Area
- Wetland

N

0   4.5   9   18
Kilometers

WGS 1984 UTM Zone 43N

Data Sources:
BODC, IHO, and IOC. 2003. GEBCO Digital Atlas (bathymetry);
and ME (administrative areas, land use and land cover).

Map I.55: Laamu Atoll, Land Use and Land Cover

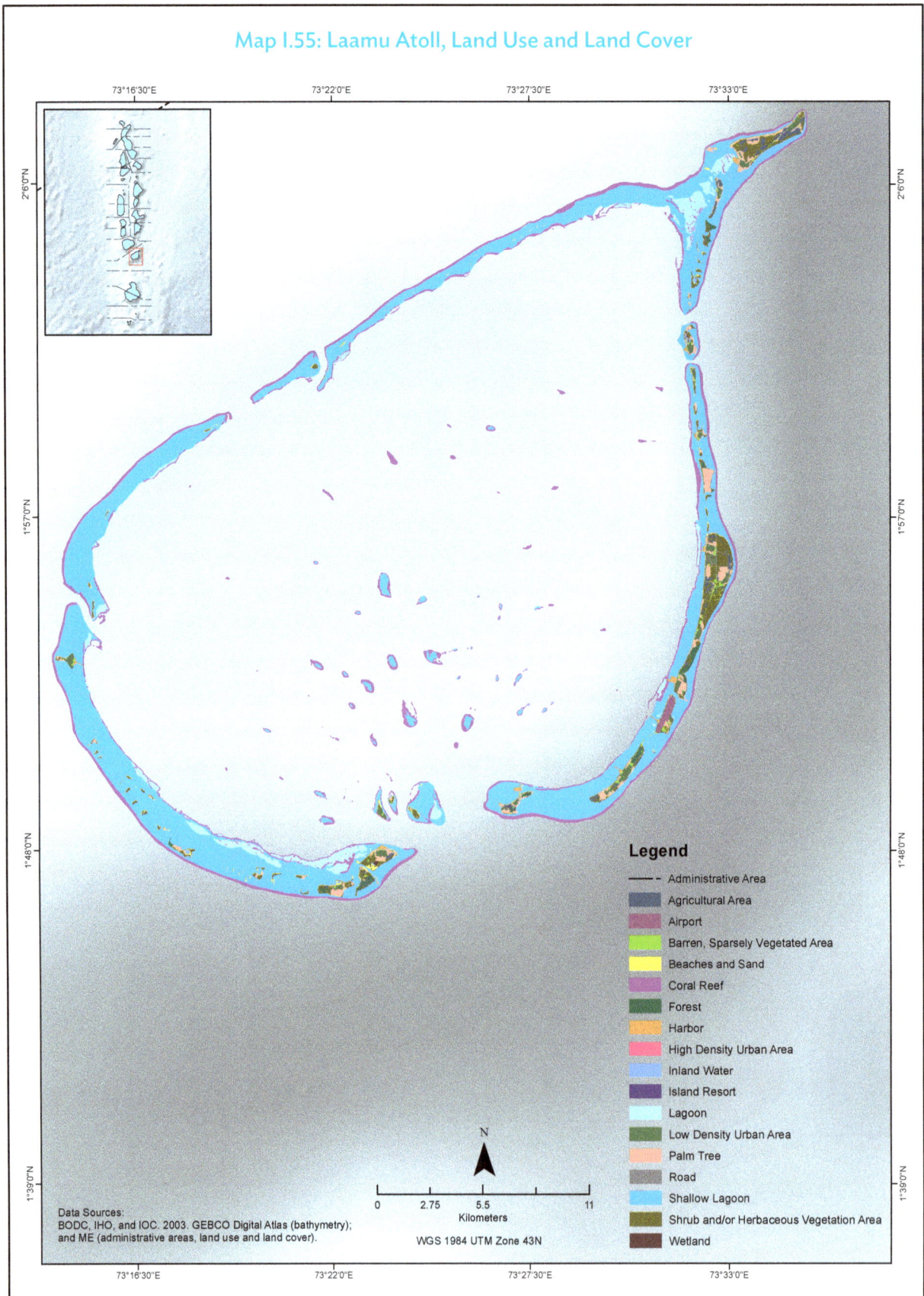

73°16'30"E   73°22'0"E   73°27'30"E   73°33'0"E

2°6'0"N

1°57'0"N

1°48'0"N

1°39'0"N

Legend

— — Administrative Area
Agricultural Area
Airport
Barren, Sparsely Vegetated Area
Beaches and Sand
Coral Reef
Forest
Harbor
High Density Urban Area
Inland Water
Island Resort
Lagoon
Low Density Urban Area
Palm Tree
Road
Shallow Lagoon
Shrub and/or Herbaceous Vegetation Area
Wetland

N

0   2.75   5.5   11
Kilometers

WGS 1984 UTM Zone 43N

Data Sources:
BODC, IHO, and IOC. 2003. GEBCO Digital Atlas (bathymetry);
and ME (administrative areas, land use and land cover).

Map I.56: Lhaviyani Atoll, Land Use and Land Cover

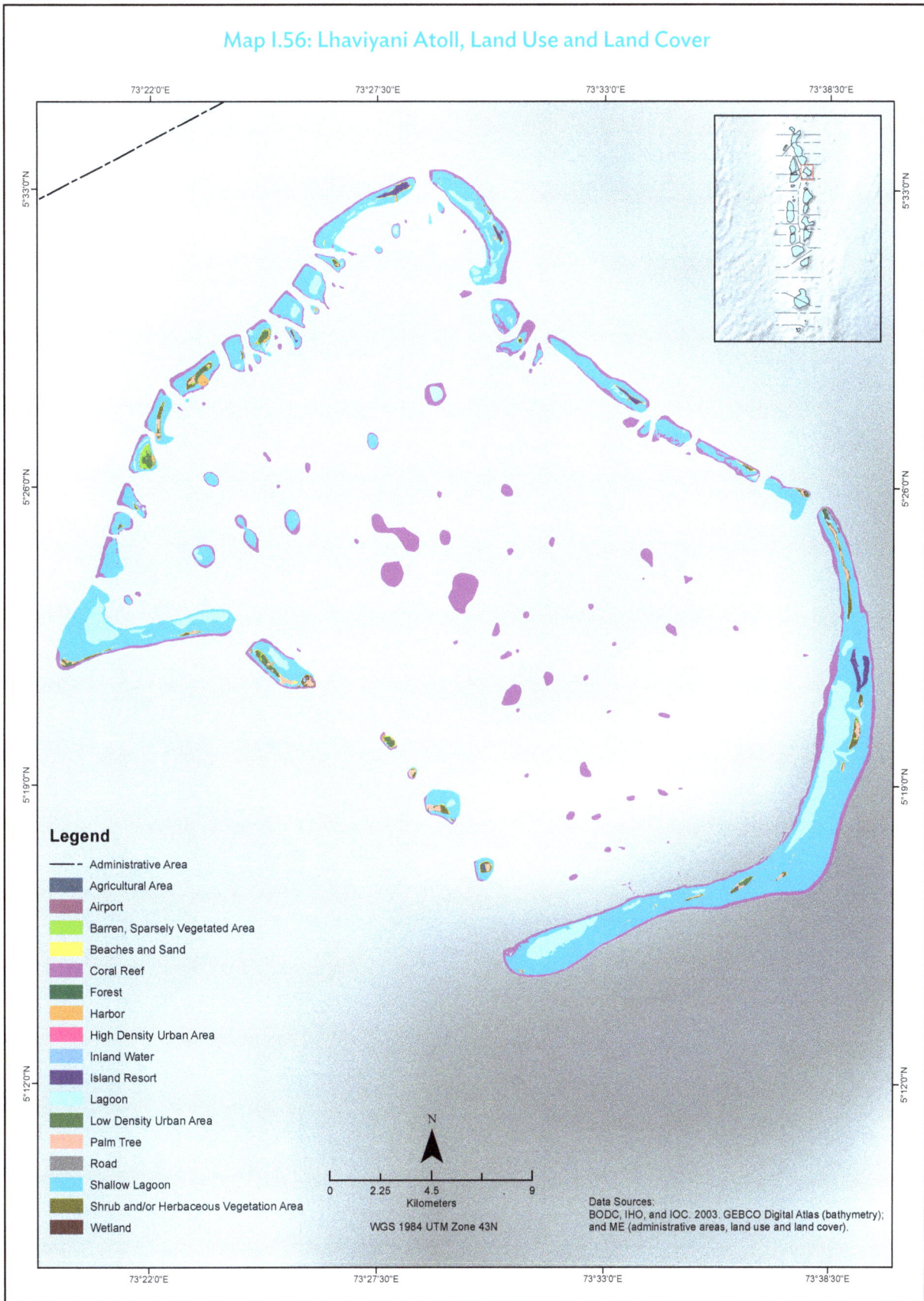

Legend

- — Administrative Area
- Agricultural Area
- Airport
- Barren, Sparsely Vegetated Area
- Beaches and Sand
- Coral Reef
- Forest
- Harbor
- High Density Urban Area
- Inland Water
- Island Resort
- Lagoon
- Low Density Urban Area
- Palm Tree
- Road
- Shallow Lagoon
- Shrub and/or Herbaceous Vegetation Area
- Wetland

0   2.25   4.5   9
Kilometers

WGS 1984 UTM Zone 43N

N

Data Sources:
BODC, IHO, and IOC. 2003. GEBCO Digital Atlas (bathymetry);
and ME (administrative areas, land use and land cover).

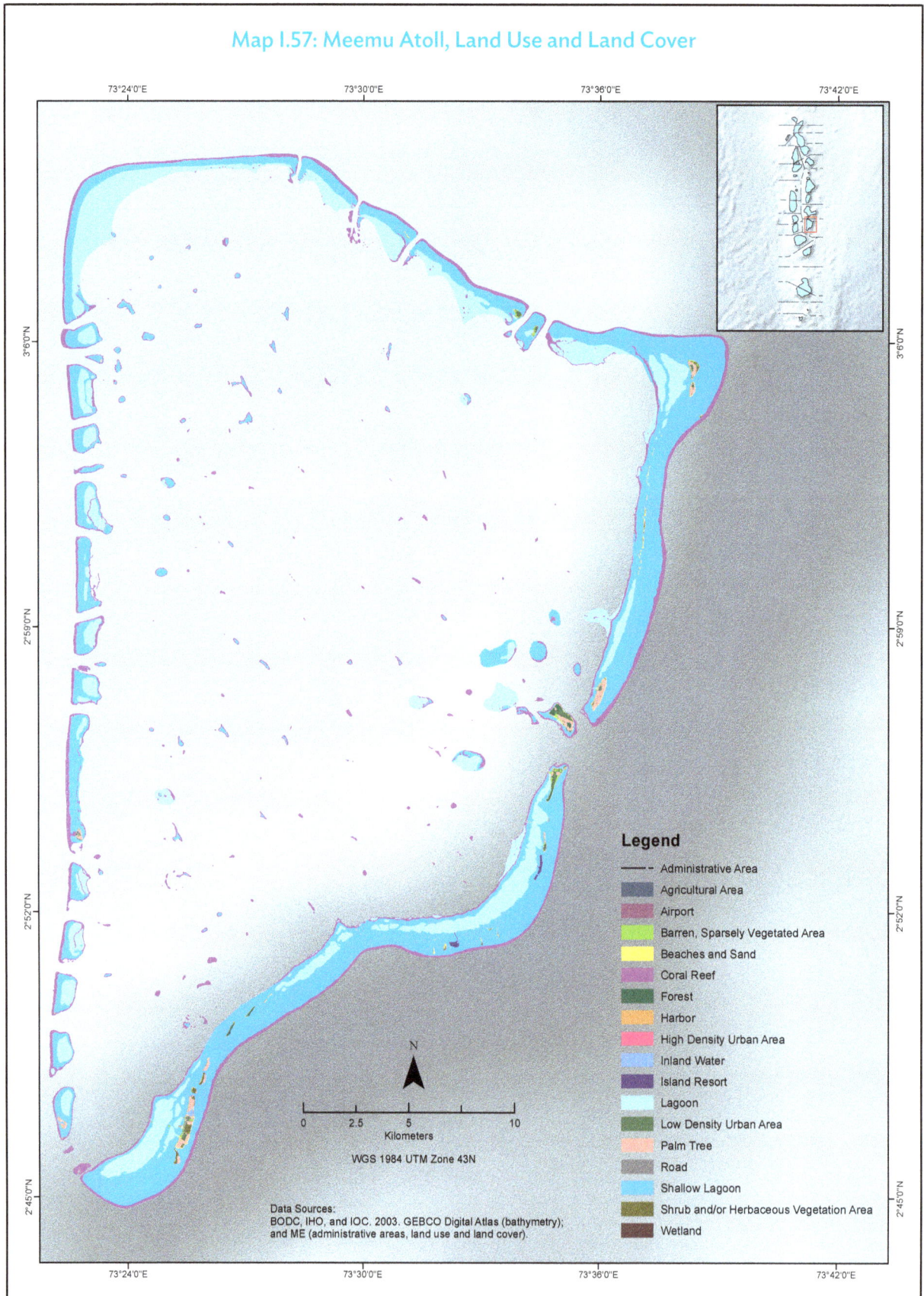

Map I.57: Meemu Atoll, Land Use and Land Cover

73°24'0"E   73°30'0"E   73°36'0"E   73°42'0"E

3°6'0"N

2°59'0"N

2°52'0"N

2°45'0"N

N

0   2.5   5   10
Kilometers

WGS 1984 UTM Zone 43N

Data Sources:
BODC, IHO, and IOC. 2003. GEBCO Digital Atlas (bathymetry);
and ME (administrative areas, land use and land cover).

Legend

- — - — Administrative Area
- Agricultural Area
- Airport
- Barren, Sparsely Vegetated Area
- Beaches and Sand
- Coral Reef
- Forest
- Harbor
- High Density Urban Area
- Inland Water
- Island Resort
- Lagoon
- Low Density Urban Area
- Palm Tree
- Road
- Shallow Lagoon
- Shrub and/or Herbaceous Vegetation Area
- Wetland

Map I.58: Noonu Atoll, Land Use and Land Cover

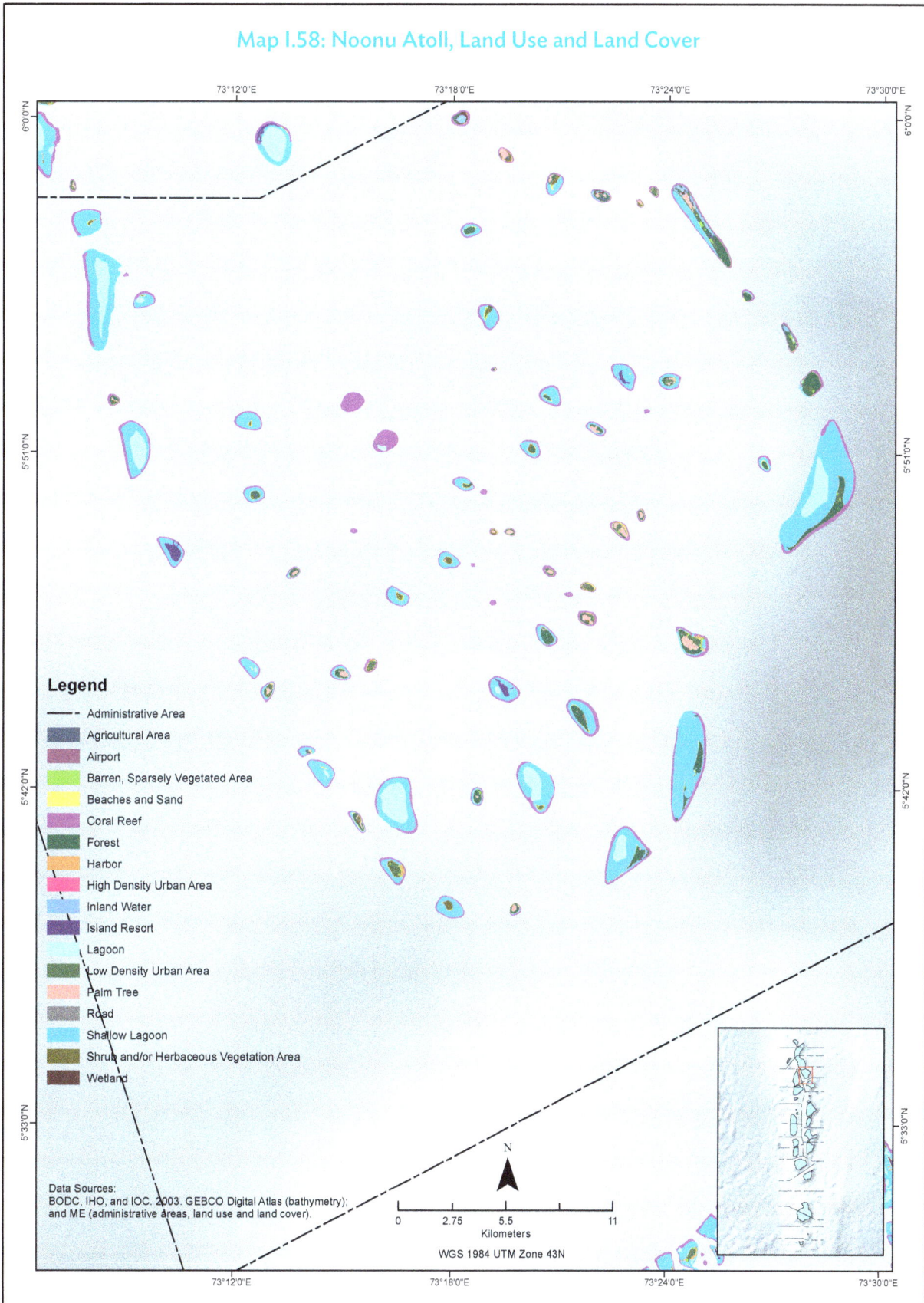

Legend

- Administrative Area
- Agricultural Area
- Airport
- Barren, Sparsely Vegetated Area
- Beaches and Sand
- Coral Reef
- Forest
- Harbor
- High Density Urban Area
- Inland Water
- Island Resort
- Lagoon
- Low Density Urban Area
- Palm Tree
- Road
- Shallow Lagoon
- Shrub and/or Herbaceous Vegetation Area
- Wetland

Data Sources:
BODC, IHO, and IOC. 2003. GEBCO Digital Atlas (bathymetry);
and ME (administrative areas, land use and land cover).

N

0   2.75   5.5   11
Kilometers

WGS 1984 UTM Zone 43N

Map I.59: North Malé Atoll, Land Use and Land Cover

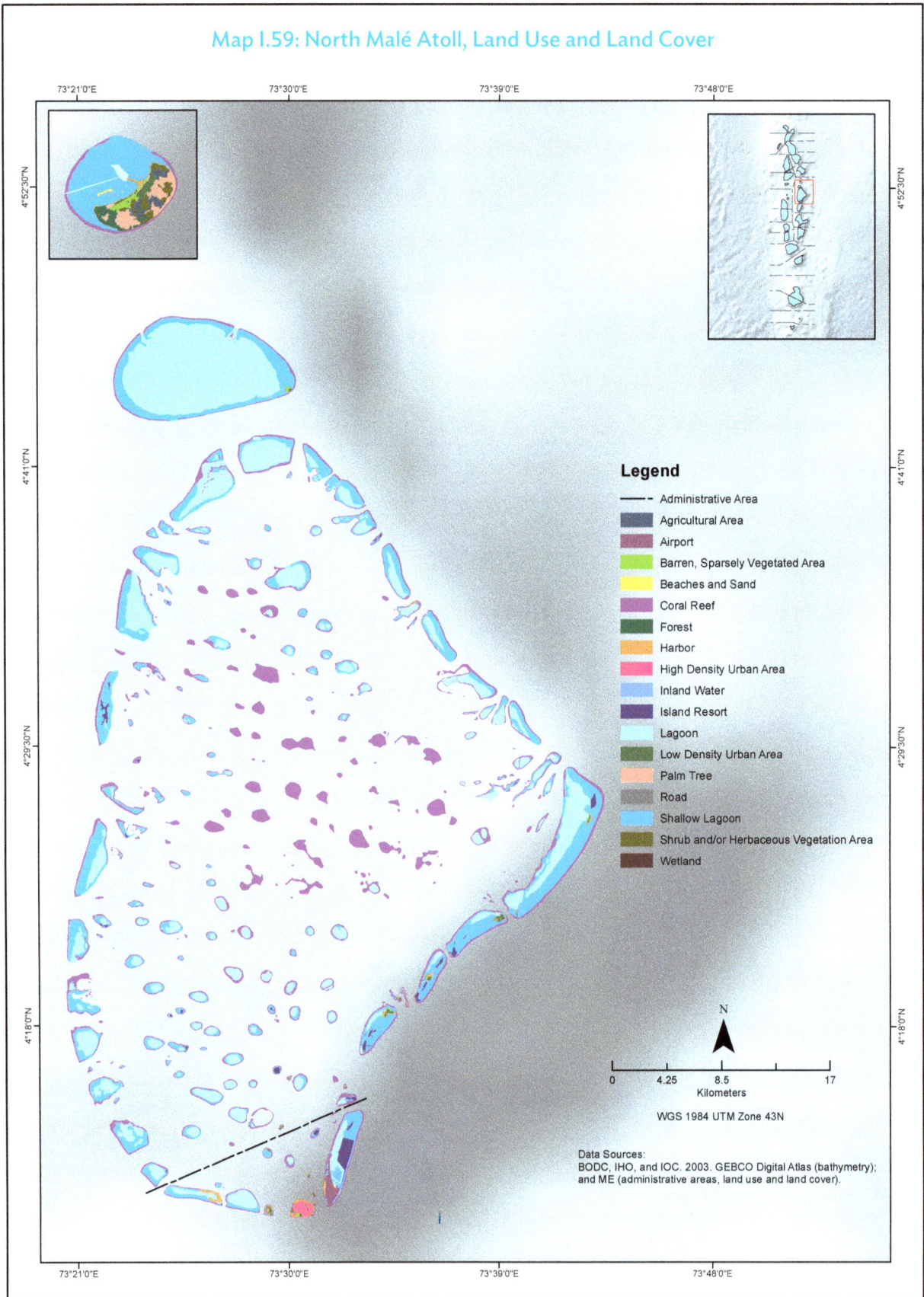

Legend

- – – – Administrative Area
- Agricultural Area
- Airport
- Barren, Sparsely Vegetated Area
- Beaches and Sand
- Coral Reef
- Forest
- Harbor
- High Density Urban Area
- Inland Water
- Island Resort
- Lagoon
- Low Density Urban Area
- Palm Tree
- Road
- Shallow Lagoon
- Shrub and/or Herbaceous Vegetation Area
- Wetland

N

0   4.25   8.5   17
Kilometers

WGS 1984 UTM Zone 43N

Data Sources:
BODC, IHO, and IOC. 2003. GEBCO Digital Atlas (bathymetry);
and ME (administrative areas, land use and land cover).

Map I.60: Raa Atoll, Land Use and Land Cover

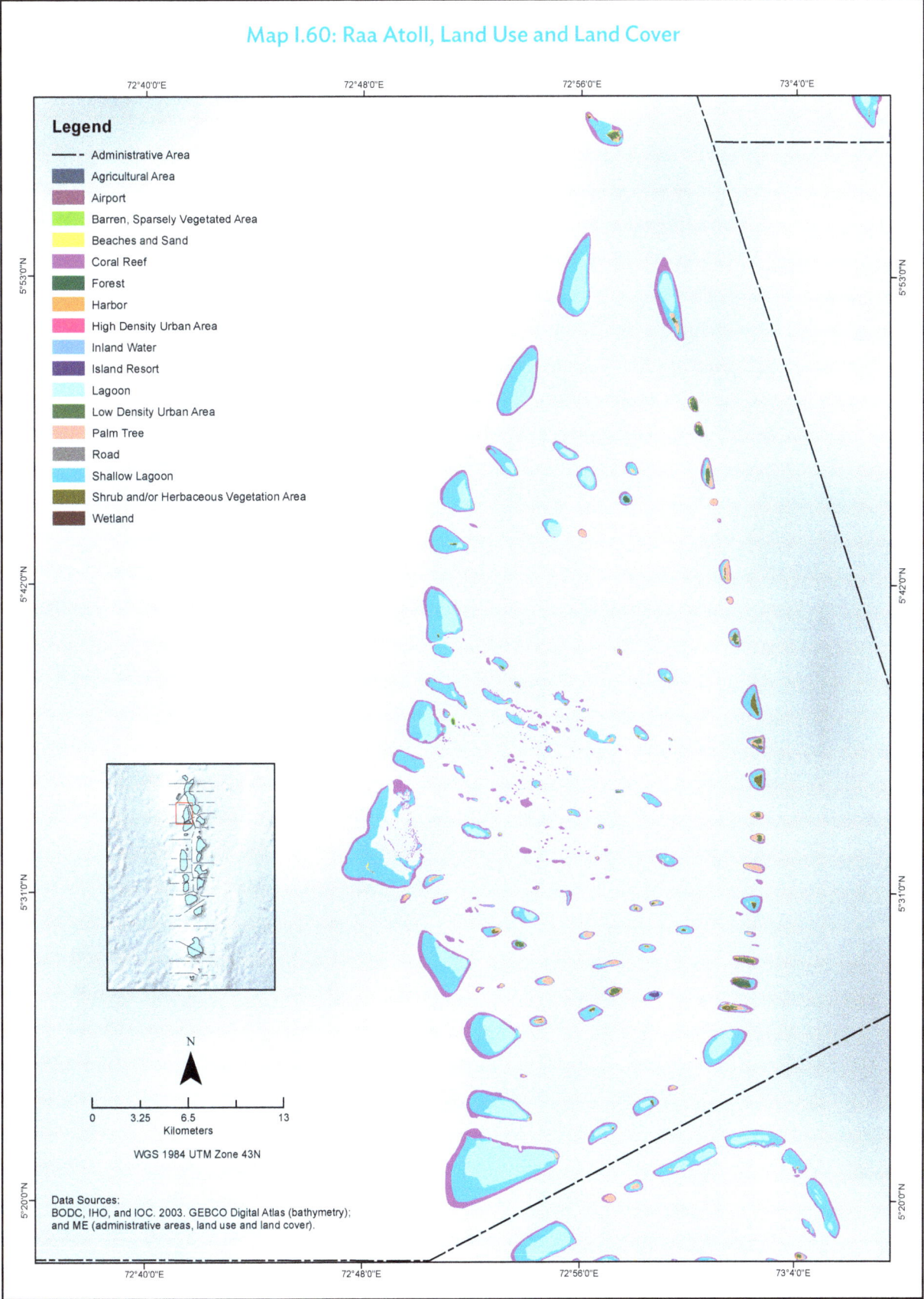

Legend

- – – – Administrative Area
- Agricultural Area
- Airport
- Barren, Sparsely Vegetated Area
- Beaches and Sand
- Coral Reef
- Forest
- Harbor
- High Density Urban Area
- Inland Water
- Island Resort
- Lagoon
- Low Density Urban Area
- Palm Tree
- Road
- Shallow Lagoon
- Shrub and/or Herbaceous Vegetation Area
- Wetland

N

0   3.25   6.5   13
Kilometers

WGS 1984 UTM Zone 43N

Data Sources:
BODC, IHO, and IOC. 2003. GEBCO Digital Atlas (bathymetry);
and ME (administrative areas, land use and land cover).

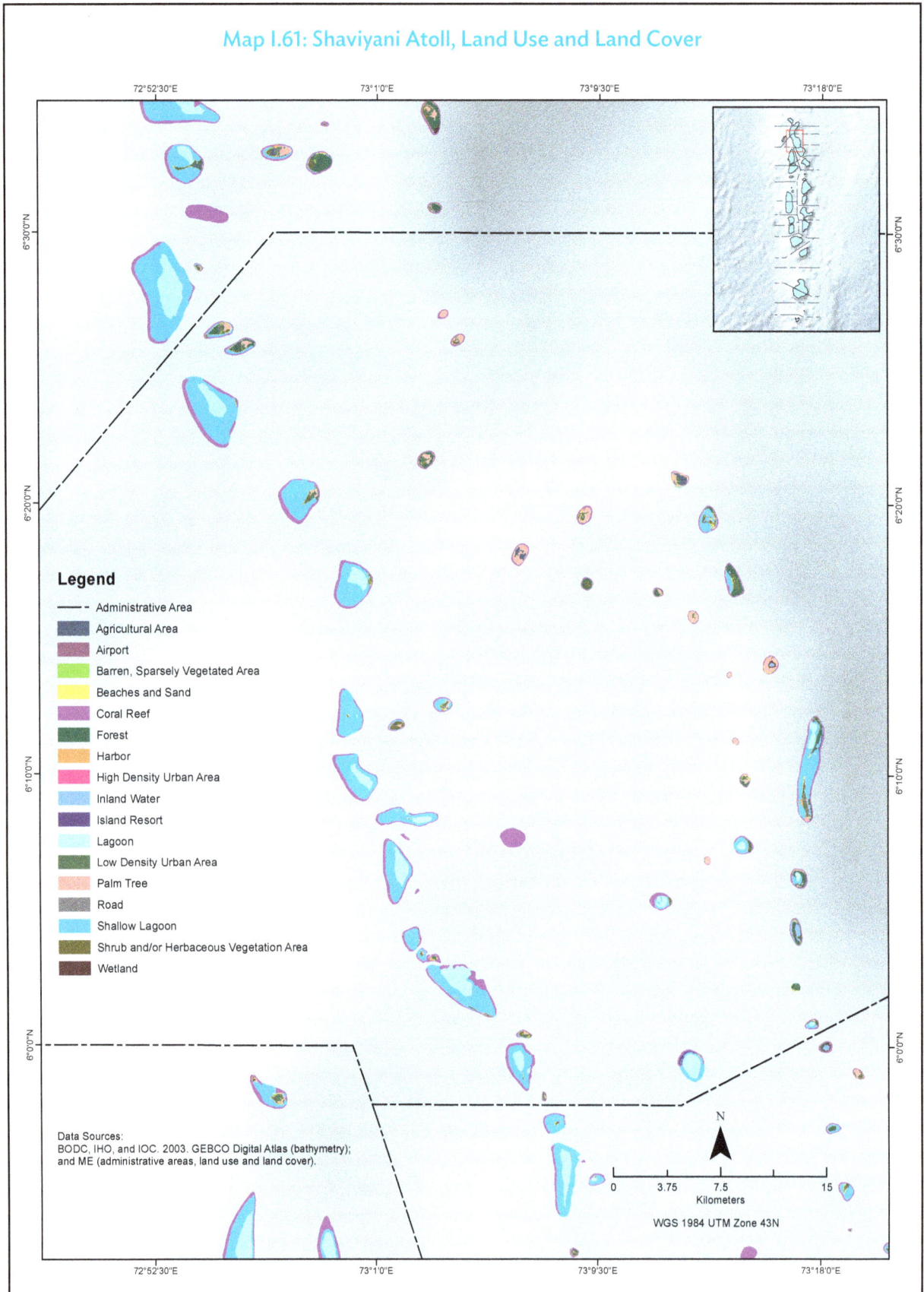

Map I.61: Shaviyani Atoll, Land Use and Land Cover

Legend

- — - Administrative Area
- Agricultural Area
- Airport
- Barren, Sparsely Vegetated Area
- Beaches and Sand
- Coral Reef
- Forest
- Harbor
- High Density Urban Area
- Inland Water
- Island Resort
- Lagoon
- Low Density Urban Area
- Palm Tree
- Road
- Shallow Lagoon
- Shrub and/or Herbaceous Vegetation Area
- Wetland

Data Sources:
BODC, IHO, and IOC. 2003. GEBCO Digital Atlas (bathymetry);
and ME (administrative areas, land use and land cover).

N

0   3.75   7.5   15
Kilometers

WGS 1984 UTM Zone 43N

Map I.62: South Malé Atoll, Land Use and Land Cover

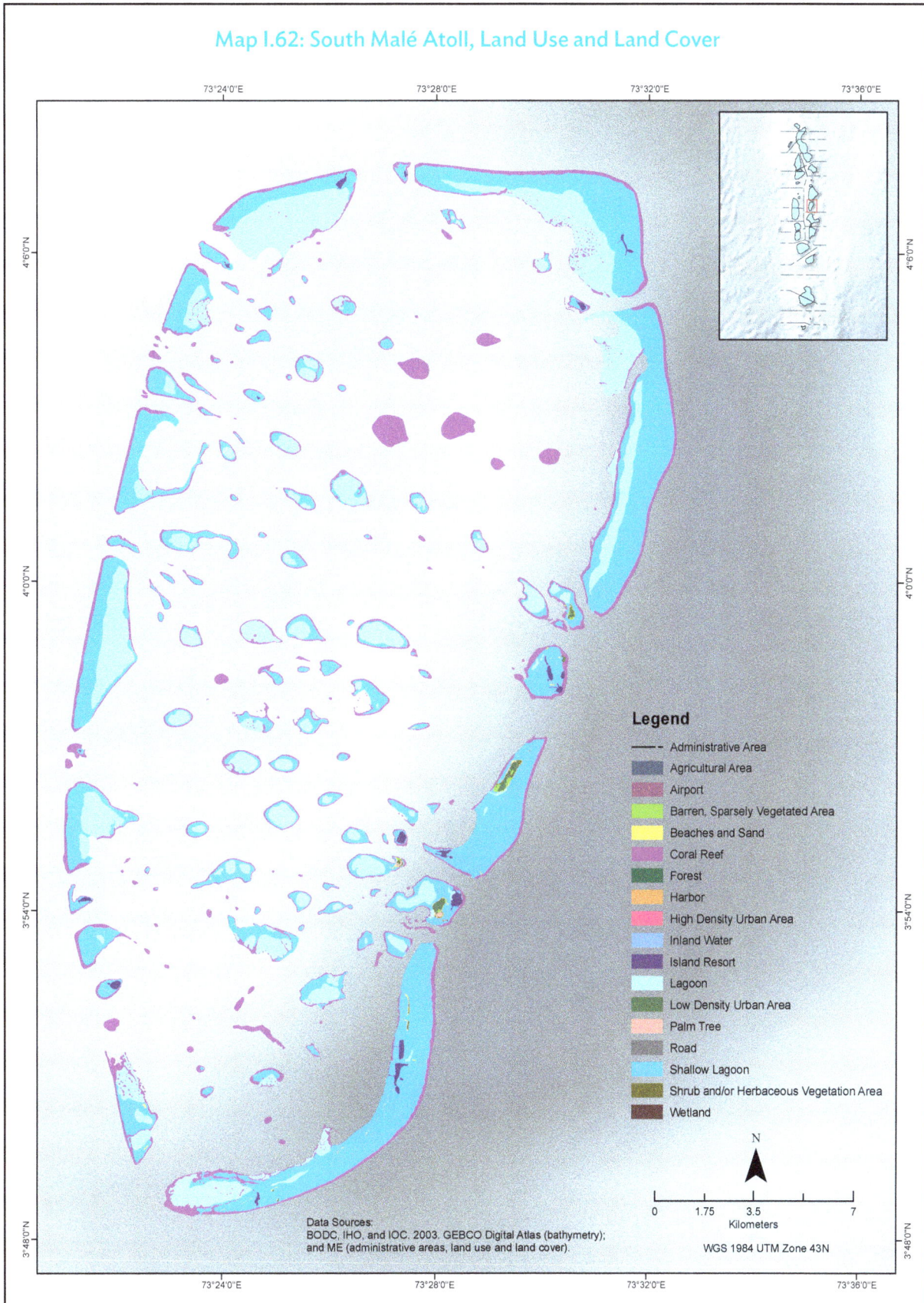

73°24'0"E   73°28'0"E   73°32'0"E   73°36'0"E

4°6'0"N

4°0'0"N

3°54'0"N

3°48'0"N

Legend

— · — Administrative Area

Agricultural Area

Airport

Barren, Sparsely Vegetated Area

Beaches and Sand

Coral Reef

Forest

Harbor

High Density Urban Area

Inland Water

Island Resort

Lagoon

Low Density Urban Area

Palm Tree

Road

Shallow Lagoon

Shrub and/or Herbaceous Vegetation Area

Wetland

N

Data Sources:
BODC, IHO, and IOC. 2003. GEBCO Digital Atlas (bathymetry);
and ME (administrative areas, land use and land cover).

0   1.75   3.5   7

Kilometers

WGS 1984 UTM Zone 43N

73°24'0"E   73°28'0"E   73°32'0"E   73°36'0"E

Map I.63: Thaa Atoll, Land Use and Land Cover

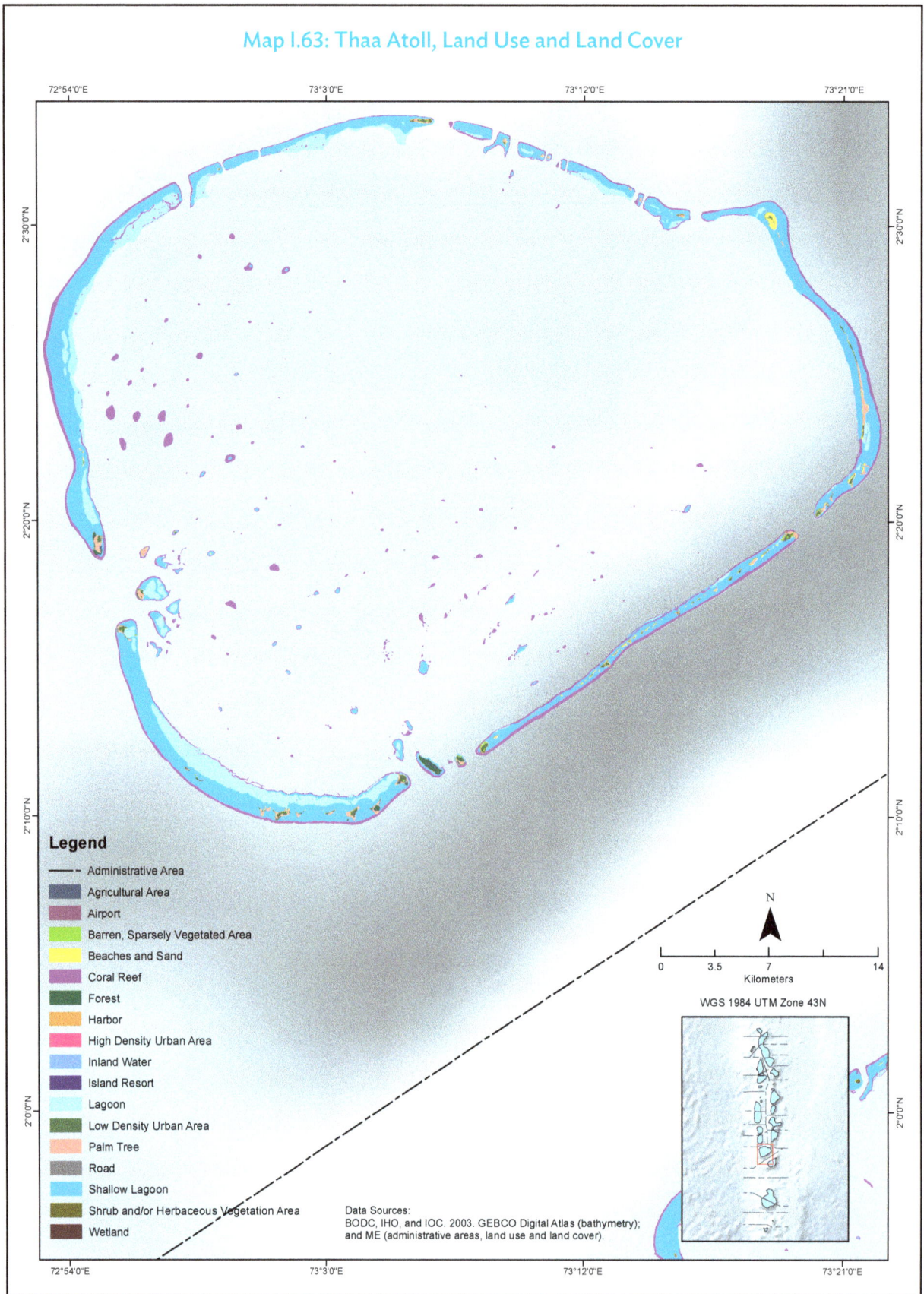

72°54'0"E 73°3'0"E 73°12'0"E 73°21'0"E

2°30'0"N
2°20'0"N
2°10'0"N
2°0'0"N

Legend

- — Administrative Area
- Agricultural Area
- Airport
- Barren, Sparsely Vegetated Area
- Beaches and Sand
- Coral Reef
- Forest
- Harbor
- High Density Urban Area
- Inland Water
- Island Resort
- Lagoon
- Low Density Urban Area
- Palm Tree
- Road
- Shallow Lagoon
- Shrub and/or Herbaceous Vegetation Area
- Wetland

N

0 3.5 7 14
Kilometers

WGS 1984 UTM Zone 43N

Data Sources:
BODC, IHO, and IOC. 2003. GEBCO Digital Atlas (bathymetry);
and ME (administrative areas, land use and land cover).

Map I.64: Vaavu Atoll, Land Use and Land Cover

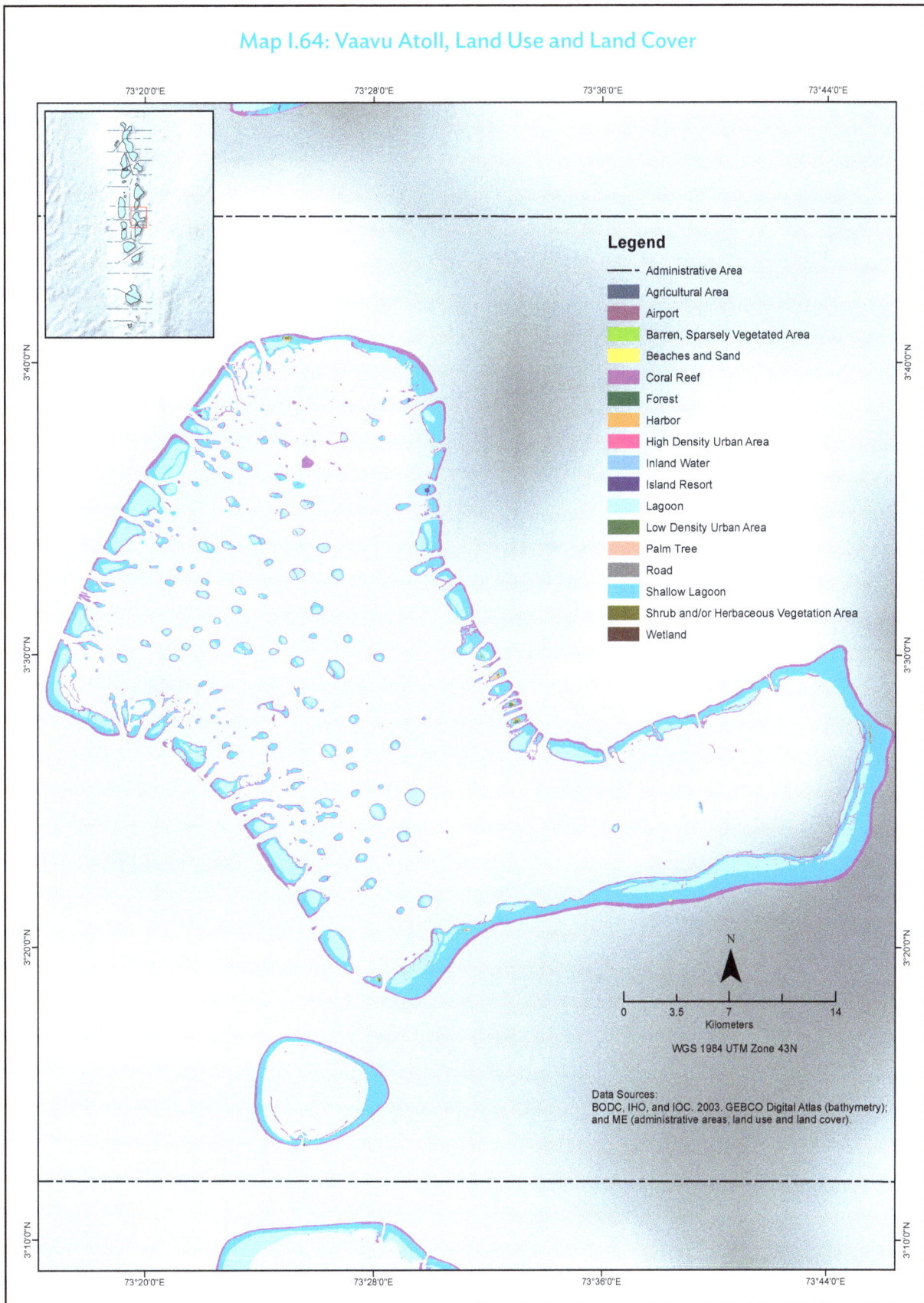

Legend

- Administrative Area
- Agricultural Area
- Airport
- Barren, Sparsely Vegetated Area
- Beaches and Sand
- Coral Reef
- Forest
- Harbor
- High Density Urban Area
- Inland Water
- Island Resort
- Lagoon
- Low Density Urban Area
- Palm Tree
- Road
- Shallow Lagoon
- Shrub and/or Herbaceous Vegetation Area
- Wetland

N

0 3.5 7 14
Kilometers
WGS 1984 UTM Zone 43N

Data Sources:
BODC, IHO, and IOC. 2003. GEBCO Digital Atlas (bathymetry); and ME (administrative areas, land use and land cover).

Map Data Sources

Government Ministries, Departments, and Agencies in Maldives
- Civil Aviation Authority
    - Airports
- Land and Survey Authority
    - Atoll capital islands
    - Cities
- Ministry of Economic Development
    - Ports
- Ministry of Environment
    - Administrative area
    - Administrative atoll
    - Island shorelines
    - Land use and land cover
    - Reef boundaries
    - Water bodies
- Ministry of National Planning and Infrastructure
    - Land reclamation

References

Asian Development Bank. 2017. *Climate Risk Screening for Mainstreaming Climate Change Adaptation into Development Activities and Policies in the Maldives.* Consultants' report. Manila (TA 8572-REG).

Ahmed, M., and S. Suphachalasai. 2014. *Assessing the Costs of Climate Change and Adaptation in South Asia.* Manila: Asian Development Bank, UK Aid.

British Oceanographic Data Centre (BODC), International Hydrographic Organisation (IHO) and the Intergovernmental Oceanographic Commission (IOC) of the United Nations Educational, Scientific and Cultural Organization. 2003. *General Bathymetric Chart of the Oceans (GEBCO) Digital Atlas.* UK: British Oceanographic Data Centre.

Khan, T., D. Quadir, T. Murty, A. Kabir, F. Aktar, and M. Sarker. 2002. Relative Sea Level Changes in Maldives and Vulnerability of Land Due to Abnormal Coastal Inundation. *Marine Geodesy* 25 (1–2). pp. 133–143.

Waheed, M., and H. Shakoor. 2015. The Impact of the Indian Ocean Tsunami on Maldives. In C. Bassard, D. W. Giles, and A. M. Howitt, eds. *Natural Disaster Management in the Asia-Pacific: Policy and Governance.* pp. 49–68. New York: Springer.